Knowledge and Technology Transfer for Plant Pathology

Plant Pathology in the 21st Century

For other titles published in this series, go to
www.springer.com/series/8169

Nigel V. Hardwick · M. Lodovica Gullino
Editors

Knowledge and Technology Transfer for Plant Pathology

Springer

Editors
Nigel V. Hardwick
The Food and Environment Research Agency
Sand Hutton
Yorkshire YO41 1LZ
UK
nigel.hardwick@fera.gsi.gov.uk

M. Lodovica Gullino
Università di Torino
Centro di Competenza per
l'innovazione in Campo
Agro-Ambientale
Via Leonardo da Vinci, 44
10095 Grugliasco TO
Italy
marialodovica.gullino@unito.it

ISBN 978-94-007-3188-2 ISBN 978-1-4020-8934-3 (eBook)
DOI 10.1007/978-1-4020-8934-3
Springer Dordrecht Heidelberg London New York

Cover illustration: Plant health clinic held in Sundarbazar, Lamjung Province, Nepal in December 2008. Photograph by E. Boa.

Printed on acid-free paper

Springer is part of Springer Science+Business Media (www.springer.com)

Preface

This book contains fuller versions of the papers and posters presented in the Knowledge and Technology Transfer and Teaching Plant Pathology sessions at the 9th International Congress of Plant Pathology held in Turin, Italy in 2008.

Communication is an essential area for plant pathologists and it is not just the publication of results in the scientific press that is important. In a world where there is a major shortage of food and where a significant amount of it is destroyed by pests and diseases before it ever reaches the consumer, it is important to provide support to those who produce the food in order to reduce the losses. Reducing crop losses not only has an impact on health, but also wealth and, therefore, the ability to survive. With an ever-increasing demand on food supplies due to increases in population, and changes in life-style associated with rising incomes in certain parts of the world, plant pathologists have a pivotal role to play in contributing to global food security.

Aspects of crop protection have lost favour with the general public because of concerns about environmental pollution and genetic modification of crops. This has had a 'knock on' effect in the recruitment and training of crop protectionist in general and a concomitant impact on courses available at universities. However, it has never been more important to train people with good communication skills and an ability to solve problems to tackle the complexities of pathogen and plant interactions.

Extension/advisory plant pathology and teaching are about relationships: the relationship between the advisor and grower and the teacher and student. It is about the building up of trust in the advice given and the enthusiasm of, and methods used by, the teacher. Extension plant pathology has been described as 'plant pathology with a human face'. It is aspects of these personal relationships that are explored in this book.

Plant pathology is an applied science – or should be. It is about solving problems in a practical way. There is no point in being able to unravel a gene sequence of, for example, *Phytophthora infestans*, if farmers do not receive advice on how this pathogen can be controlled economically and in their own, and the best interests of the community. Farmers are not generally interested in the nuances of the cause of a disease; they want solutions that are easy to apply and are economic. They may be content to do what they are advised or they may wish to see the advice demonstrated. Each farmer is different. They have their own aspirations. A farmer on a rented farm may need to maximise profit in order to pay the rent, an owner occupier may forgo profit to have a more relaxing life style. Farmers in less developed countries need to protect the crops they grow for their very survival. So the aspirations of farmers vary according to circumstances. Advisers/extension workers at the field gate have to bear all this in mind in their interpretation of the advice they give, for what will suit one farmer may not suit another.

Plant pathologists have always taken the lead in establishing plant clinics, a major resource for identifying crop problems. While entomologists, agronomists and soil scientists play their part, in my experience, it is generally left to the plant pathologist to interpret the result of a diagnosis. It is here that vital training and teaching plays its part. The understanding of the basic biology, the interactions between the disease, its host and context are important but so is experience. Plant pathologists also seem imbued with a sixth sense. They often have to make recommendations and provide advice, not only on the basis of the science but on an understanding, gained over several years of experience of the level of disease present, the growth stage of the crop, the general conditions affecting the crop, local topography and likely future weather. Not forgetting the market and general aspirations of the farmer/grower.

The teaching of those who are to become advisors and technicians in the control of crop diseases and the means of working with farmers and growers to increase production and the benefits this will accrue, are all explored in the following chapters.

N. V. Hardwick

Contents

Contributors

E. Boa
Global Plant Clinic, CABI E-UK, Bakeham Lane, Egham, Surrey TW20 9TY, UK
e.boa@cabi.org

C. Bragard
Unité de Phytopathologie, Université catholique de Louvain, Croix du Sud 2/3, B-1348 Louvain-la-Neuve, Belgium

N. Capodagli
Huayimeng Environment, Science & Technology Consulting Co., Ltd, Beijing, China

D.M. Eastburn
Department of Crop Sciences, University of Illinois, Urbana, IL, USA
eastburn@illinois.edu

R. Daniel
Faculty of Agriculture, Food and Natural Resources, The University of Sydney, Sydney, Australia
r.daniel@usyd.edu.au

C.J. D'Arcy
Department of Crop Sciences, University of Illinois, Urbana, IL, USA

A. Garibaldi
Centre of Competence for the Innovation in the agro-environmental sector (AGROINNOVA), University of Torino, Via Leonardo da Vinci 44, 10095 Grugliasco (Torino), Italy

U. Gisi
Syngenta Crop Protection, Biology Research Center, Schaffhauserstrasse, CH-4332 Stein, Switzerland and Institute of Botany, University of Basel, Hebelstrasse, CH-4056 Basel, Switzerland
ulrich.gisi@syngenta.com

D.I. Guest
Faculty of Agriculture, Food and Natural Resources, The University of Sydney, Sydney 2006, Australia
d.guest@usyd.edu.au

M.L. Gullino
Centre of Competence for the Innovation in the agro-environmental sector (AGROINNOVA), University of Torino, Via Leonardo da Vinci 44, 10095 Grugliasco (Torino), Italy
marialodovica.gullino@unito.it

J.K. Konam
Plant Health Team, Land Resources Division, South Pacific Commission, Private Mail Bag, SUVA, Fiji Islands
JohnK@spc.int

U. Krauss
CAB International (CABI), Gordon Street, Curepe, Trinidad and Tobago
u.krauss@cabi.org

A. Legrève
Unité de Phytopathologie, Université catholique de Louvain, Croix du Sud 2/3, B-1348 Louvain-la-Neuve, Belgium

H. Maraite
Unité de Phytopathologie, Université Catholique de Louvain, Croix du Sud 2/3, B-1348 Louvain-la-Neuve, Belgium
henri.maraite@uclouvain.be

Y. Namaliu
PNG Cocoa and Coconut Institute, Stewart Research Station, PO Box 642, Madang, Papua New Guinea
yaknamaliu@yahoo.com.au

S.L.J. Page
CAB International (CABI), Bakeham Lane, Egham, Surrey TW20 9TY, UK
s.page@cabi.org

M. Pugliese
Centre of Competence for the Innovation in the Agro-Environmental Sector (AGROINNOVA), University of Torino, Via Leonardo da Vinci 44, 10095 Grugliasco (Torino), Italy

J.P. Stack
Department of Plant Pathology, 4024 Throckmorton Plant Sciences Center, Manhattan, Kansas, 66506, USA
jstack@ksu.edu

P. Vincelli
Department of Plant Pathology, University of Kentucky, 207 Plant Science Building, Lexington, KY, 40546-0312
pvincell@uky.edu

J.G.M. Vos
CAB International (CABI), Kastanjelaan 5, 3833 AN Leusden, the Netherlands
j.vos@cabi.org

Chapter 1
Plant Healthcare for Poor Farmers Around the World: Gathering Demand and Innovative Responses

E. Boa

1.1 Introduction

The first plant health clinics[1] began in Bolivia in 2003 with the help of the Global Plant Clinic (GPC). Six years on there are nine regular clinics operated by the two original clinic organisations, CIAT Santa Cruz[2] and PROINPA, with the Universidad Mayor de San Simón (UMSS) joining the scheme in 2006. The clinics in Bolivia have inspired people in other countries to run their own clinics and, by early 2009, there were schemes in ten countries across Latin America, Africa and Asia, including Bolivia, giving a total of 81 regular clinics who receive support from the GPC (Table 1.1, Fig. 1.1). The number of regular clinics varies but is remarkably stable given uncertain funding and weak support from governments to agriculture.

The origins of the GPC and the beginnings of the plant health clinics in Bolivia are described by Boa (2009). The purpose of this chapter is to draw upon experiences from a larger group of clinics and country schemes in order to explain how clinics function; the role of the GPC in relation to plant doctors and the organisations they work for; and to present selected results from clinics.

As the more advanced schemes in Nicaragua (Danielsen and Fernández 2008), Bangladesh and Bolivia have grown so their successes have encouraged the development of networks that link clinics first to each other and then groups of clinics to other resources, for example diagnostic laboratories. The Bangladesh and Bolivia schemes have been operating for longer than Nicaragua and have generated many new ideas

[1] Plant health clinic is the preferred title though this is shortened to plant clinic or simply clinic. In Bolivia the clinics are called Postas para Plantas ('plant posts'), by analogy with human health clinics (Posta de Salud) and Puestos para Plantas in Nicaragua. Other variants include mobile plant clinic and rural plant clinic.

[2] An agricultural institute belonging to the Department of Santa Cruz in Bolivia and distinct from CIAT in Colombia, an international research institute and member of the CGIAR.

E. Boa (✉)
Global Plant Clinic, CABI E-UK, Bakeham Lane, Egham, Surrey TW20 9TY, UK
e-mail: e.boa@cabi.org

N.V. Hardwick and M.L. Gullino (eds.), *Knowledge and Technology Transfer for Plant Pathology*, Plant Pathology in the 21st Century 4,
DOI 10.1007/978-1-4020-8934-3_1, © Springer Science+Business Media B.V. 2010

Table 1.1 Plant health clinics supported by the Global Plant Clinic from 2003 up to June 2009

Country	Year started	No. of clinics	Frequency
Bangladesh	2004	19	Regular
Benin	2006	(3)	Pilot
Bolivia	2003	9	Regular
Cameroon	2007	2	Pilot
Colombia	2006	(2)	Pilot
Côte d'Ivoire	2006	(1)	Pilot
Cuba	2005	(2)	Pilot
DR Congo	2006	9	Regular
Guinea	2006	(1)	Pilot
India	2006	2	Regular
Indonesia	2007	(2)	Pilot
Kenya	2005	(4)	Pilot
Mali	2008	(1)	Pilot
Nepal	2009	3	Regular
Nicaragua[a]	2005	18	Regular
Pakistan	2009	(2)	Pilot
Peru	2008	(2)	Pilot
Rwanda	2006	1 and (3)	Regular from 2009
Sierra Leone	2006	13	Regular since 2008
Sri Lanka	2009	(1)	Pilot
Uganda	2005	4	Regular since 2006
Vietnam[b]	2007	1	Regular
Total		105	81 Regular, 24 Pilots

Number of pilot clinics are shown in brackets
[a]The number of clinics has fluctuated in Nicaragua since 2007 but new ones will start in 2009
[b]In addition to a regular clinic at SOFRI, plant doctors have held more than 11 mobile clinics up to 200 km from the institute headquarters in My Tho

and results which have not yet been published. However, reports are available at www.reasearch4development.info under the GPC section several articles have been published (e.g. Bentley et al. 2007; Kelly et al. 2008) and a major paper on Bolivia has recently been published (Bentley et al. 2009a).

The activities of the GPC extend beyond establishing and supporting plant clinics. The initial aim of the GPC was to work 'more closely with farmers' and organisations who supported them. We began with short training courses on 'field diagnosis' (recognition and interpretation of symptoms) which have since expanded to address broader needs; agriculturists and farmers required more than better diagnostics. As our engagement with these groups has increased so has our understanding of how to help improve plant health systems and wider aspects of extension. I will include a brief review of a new extension method, Going Public (Bentley et al. 2003), plant disease surveillance and new disease records and ethnopathology studies (Bentley et al. 2009b). These are three examples of innovative actions and responses that originated through interactions between the GPC, farmers, and staff who help run plant clinics.

In his classic book about plant diseases, 'The Advance of the Fungi', Large (1940) talks about plant health systems and plant doctors. This was a natural consequence

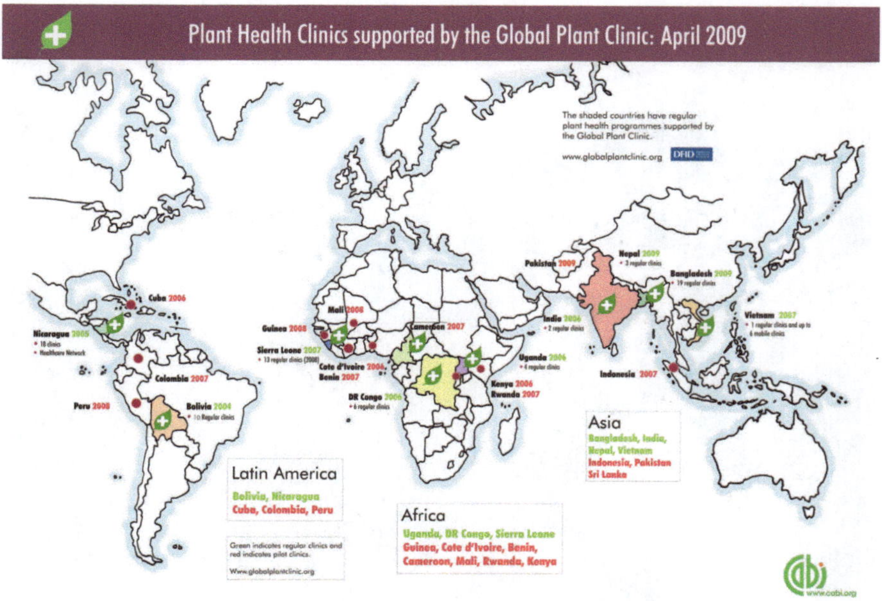

Fig. 1.1 Plant clinics supported by the Global Plant Clinic up to April 2009

at the time of drawing analogies between plant health and human health, a connection which has since faded. Anderson (2007) notes a renewed importance and interest in extension and advisory services and the unequal influence of research and researchers able to exert a stronger influence on agricultural agendas.

Despite a modern concern for extension, an historical neglect has led to a lack of innovation in advisory methods. The time has come to rethink how farmers' needs are best served, particularly in developing countries, where agriculture is the most important business for hundreds of millions of poor people. Yet farmers in developing countries are poorly served by extension services. Despite the huge numbers of extension workers employed by governments, accountability to farmers is low (Anderson 2007) and farming communities are widely dispersed. Even the best extension workers struggle to meet all their assigned clientele and farmers have few reliable and consistent means for expressing their demands.

Plant health clinics are not a new idea and the United States and other developed countries have effective plant health systems that serve farmers well and have done so for many years (Campbell et al. 1999). In the context of developing countries, however, plant clinics are a 'new' method for farmers and their communities. A majority who visit the clinics say this is the first time they have been able ask for help with a plant health problem (Fig. 1.2).

Plant clinics are a new opportunity to create effective plant health systems in developing countries and to deliver plant healthcare to more farmers. Healthcare is an important concept since it encompasses treatment and management of illness

Fig. 1.2 Plant health clinic held in Sundarbazar, Lamjung Province, Nepal in December 2008. Farmers bring in diseased samples which the plant doctors, supported by World Vision International Nepal, examine and give recommendations for control. Many women farmers attended who had travelled for several hours to get to the clinic

and the preservation of good health. It is a familiar concept in human and animal health yet receives little mention for plants. Plant healthcare is central to the GPC's mission since it encourages us and our partners to consider all aspects of how plants are grown and to seek useful solutions for all types of problems, not just pests and diseases. Of course this also increases the challenges for plant doctors in gaining knowledge and skills necessary to diagnose problems and find these solutions, but these are challenges that need to be firmly addressed if our interventions are to have an impact on farmers' livelihoods.

Richards (1985) was one of the first to highlight the gap in agriculture between science and development, based largely on his experiences in Sierra Leone – coincidentally a country that began plant clinics in 2008. While this gap has been reduced over the last 30 years through greater involvement of farmers in research and extension, and IPM farmer field schools have contributed much (Van de Berg and Jiggins 2007), a recent review of the impact of IPM extension emphasises that there is still much to do (Bentley 2009).

Plant health clinics alone will not cure systemic weaknesses in extension services or low adoption of technologies, but they do offer new advantages: they are demand-led; they provide regular and reliable advice to more farmers; the plant doctors running them are accountable to the clinic users. There are also new collaborations to be made between clinics and farmer field schools, as Bentley (2009) has suggested: clinics respond directly to farmer problems; farmer field schools, can carry out locally adaptive research to find new or better solutions to problems.

The long term aim of the GPC is to create plant health systems within which clinics form a firm foundation for wider interventions and responses to farmer needs. The first step is to strengthen links with diagnostic laboratories and plant diagnostic networks (Miller et al. 2009; Smith et al. 2008). There are encouraging signs in Bolivia and Nicaragua of farmers sending in samples and gaining access to diagnostic services for the first time facilitated by plant clinics. Establishing clinics remains the first priority of the GPC, followed by monitoring of progress and quality control of services offered and the impact these have on farmer livelihoods. The many aspects of establishing and operating clinics are described by Danielsen and Fernández (2008) for Nicaragua.

1.2 Plant Health Clinics

Plant health clinics have several unique features: they accept any crop and any problem; they are run independently by organisations using existing facilities and personnel; they are open to all farmers. Plant health clinics are run in public places to increase visibility and access to farmers. The most popular venue is a market place, although a variety of other places are used, including agrochemical supply shops, schools, the offices of farmer co-operatives, village stalls. In Vietnam, SOFRI runs a clinic at their main research institute, which farmers regularly visit. They also hold mobile clinics up to 200 km away in order to expand their outreach.

Clinics are run by staff (plant doctors) belonging to many different types of organisations (Table 1.2) and clinic operators (Table 1.3). The success of clinics

Table 1.2 Main categories of clinic organisations who work with the GPC

Category	Types
Community-based	Local and international development NGOs, farmer associations, co-operatives and similar. Work directly with farm families
Local authorities	Mayoralities, municipalities, district authorities and so on. Bodies with a local or regional remit
Research institutes	Science and technology, mostly applied research. Includes local institutes as well as international
Ministries of Agriculture	Key feature is public funding disbursed on a national scale. Includes phytosanitary inspectors and agricultural officers ('extension') and diagnostic labs (Nicaragua). Some extension is funded by region
Universities	Teaching (main role) and research. Some also run diagnostic laboratories
Companies	Privately owned

Table 1.3 Summary of organisations which run or support clinics in 19 countries

Country	1 Community based	2 Research institutes	3 Minstries of Agriculture	4 Local authorities	5 Universities	6 Companies	Total	Clinic status
Bangladesh	2	1		1		2	6	Regular
Benin		1					1	Pilots
Bolivia		2		2	1		5	Regular
Cameroon		1					1	Pilots
Colombia					2		2	Pilots
Cote d'Ivoire					1		1	Pilot
Cuba		2					2	Pilots
DR Congo					1	1	2	Regular
India					2	1	3	Regular
Indonesia		1			1		2	Pilots
Kenya		1	1				2	Pilots
Mali		1					1	Pilot
Nepal	3		1				4	Regular
Nicaragua	9	5	1		3		18	Regular
Pakistan			1				1	Pilot
Peru	3	1	1				5	Pilot
Rwanda		1			1		2	Regular
Sierra Leone			1				1	Regular
Sri Lanka		1					1	Pilot
Uganda	3	1	1				5	Regular
Vietnam		1					1	Regular
All	20	20	7	3	12	4	66	

depends on the commitment of organisations as well as their staff. When clinics started in 2003 in Bolivia, the GPC already had well-established working relationships with enthusiastic staff from CIAT Santa Cruz and PROINPA, who then helped to establish clinics and keep them running. There are currently nine clinics in Bolivia.

All clinics operate in a similar manner. They are usually run by agricultural officers or technicians who take on the role of 'plant doctor' during weekly sessions which typically last for 2–3 h. The clinics consist of a table and chairs, shaded from the sun but clearly visible to passers by. A banner or prominent sign helps to attract people. These are the broad features of a plant health clinic: the location and timing depends on those conditions which attract the most clinic users. There are set instructions, however, on how to record information, using a register for the query and a prescription pad for writing a recommendation. Clinic data are vital to understanding demand and improving responses to problems. Recording the data so others can review it and benefit from the information is a major challenge, as is ensuring consistency and accuracy. But the data are also hugely revealing, showing which problems matter most to farmers and the ability of plant doctors to make helpful recommendations.

Clinics are also advertised on local radio, in newspapers, on television and by other ingenious means. In Bangladesh, the clinics are announced through a portable public address system carried through a village on a bicycle rickshaw, a method known as 'miking'. In Nepal, leaflets announcing a plant clinic were released from a car window en route to the clinic.

Farmers are requested to bring samples for plant doctors to examine, although clinics also accept many problems that lack supporting plant material. The quality of samples varies from well-selected and fresh plant parts, to dead branches, rotten plants and one or two isolated leaves. Gathering requests is a demanding job and one of the first tasks of the plant doctors is to encourage farmers to bring in good quality samples.

Clinics were started in Nicaragua in 2005 after identifying individuals keen to run them. However, individual enthusiasm is not enough if employers are unable or willing to allow the plant doctor to hold the clinic. The first clinic in Estelí, Nicaragua was run consistently for 3 years by a highly committed individual staff member, yet when a new project came along the plant doctor running the clinic was moved to a new place of work and the clinic ceased. In 2008 the GPC was invited to help establish plant clinics by World Vision International Nepal (WVIN). The initial contact with the GPC was made through a young and dynamic employee of WVIN based in Lamjung province, who had read an article about plant clinics in Nicaragua (Bentley et al. 2007). The discussions concerning the clinics involved employers of staff from the outset. Senior WVIN staff asked questions about how clinics would operate and the roles of plant doctors. They came to see a pilot clinic in operation. The Nepal clinics successfully started in the first half of 2009 with the positive endorsement of WVIN management.

As the clinics became better known and more people read about them (e.g. Bentley and Boa 2004; Bentley et al. 2007), the GPC learnt of new people and organisations interested in the idea. The plant clinics in the Democratic Republic of the Congo came about through a consultancy visit for a cacao disease in North Kivu in 2004. We met Professor Ndungo Vigheri from the Catholic University of Graben and helped him start plant clinics in 2006.

The plant clinics in Sierra Leone began in 2008. The results from other country schemes and a successful pilot clinic held in Freetown in 2006 showed what the clinics could achieve and convinced the Government to fund 13 districts, one in each district. The clinics in Bangladesh began in 2004 through already established contacts with the Rural Development Academy in Bogra and the NGO Agricultural Advisory Society. A third organisation, Shushilan, joined in 2006 (Kelly et al. 2008). In India, contact was made with GB Pant University of Agriculture and Technology and they run two clinics.

The regular clinic schemes all began with one or more pilots. Those attending received first-hand experience of how a clinic operated, saw more clearly what they could achieve and learnt some of the challenges. The GPC shared the experiences of these pilot clinics through illustrated reports which contained short narratives and a few key messages. We also published photosheets of pilot clinics (up to six photos and short captions on an A3 sheet), videos on YouTube and newsletter articles (e.g. Bentley et al. 2007).

Advocacy has been a key part of the GPC strategy to establish more clinics. Two 15 min videos were produced by CountryWise Communications for Nicaragua and Bangladesh, and they have had a major impact in stimulating new interest as well as showing how clinics link to other organisations and project activities.

From 2003 until April 2009, 24 pilot plant clinics were held in 12 countries which have yet to establish regular schemes (Table 1.1). In Indonesia and Peru we were unable to identify a suitable clinic organisation in the short time available; in Cuba agriculture is closely supervised and running clinics requires official sanction at the highest level. In some countries (particularly Africa) new activities ostensibly depend on earmarked funding from projects. Attitudes are changing however, influenced by the results that clinics achieve. GPC advocacy is an important stratagem in effecting this change.

1.3 Plant Doctors

Unlike human health, the term 'doctor' is not a protected profession in agriculture; one can use the title 'plant doctor' regardless of qualifications or accreditation. We use the title plant doctor to describe a person who diagnoses plant health problems and offers advice on how to manage them. Most plant doctors are agronomists and many have years of experience of local agriculture and working with farmers (Fig. 1.3). Many do a much better job than they are given credit for (Boa 2009).

Plant doctor does not mean that the practitioner has equal knowledge of all types of problems, hence the importance of linking clinics to other sources of technical expertise. The GPC runs several short courses on 'how to become a plant doctor'. These include modules on field diagnosis and running a clinic, plant

Fig. 1.3 Yamileth Calderón, agriculturist and plant doctor, works for an NGO and held a Puesto para Plantas in Estelí, Nicaragua very Friday. She used her extensive knowledge of local agriculture to figure out recommendations that are suitable for local farmers

healthcare and extension messages. Over 40 courses have been run since 2002 and an estimated 800 people have attended at least one course in over 20 countries. As the number of practising plant doctors increase and clinics become a routine part of extension, professional qualifications and regular testing of knowledge and skills will be required.

In Nicaragua, the GPC trained trainers to teach the courses and in Bolivia staff from existing clinics are teaching new plant doctors. New modules have been added since the basic curriculum on 'how to become a plant doctor' was finalised in Nicaragua in 2007. The most important new module is 'Monitoring progress and quality' of clinic operations (Danielsen and Kelly 2009). Clinic performance needs to be monitored for consistently and reliability, and the quality of diagnosis and advice reviewed regularly to ensure that farmers receive the best advice. Monitoring quality is part of the increased accountability that plant doctors experience in providing face to face previous advice to farmers. Farmers return to clinics and comment on previous advice they have received. Plant doctors need advice on unknown problems and want to learn how to improve the services they offer.

Accountability helps to improve the overall quality of service of clinics. Some problems need further investigation and when a farmer returns for the results the plant doctors must be ready to explain what they have found out since the original consultation. Quality is about timeliness as well as accuracy of a diagnosis.

Selected clinic data from Uganda clinics (Table 1.4) show the range of crops that farmers bring (Table 1.5). The range of problems presented in clinics can

Table 1.4 Uganda – plant health queries and crops by rank from three clinics, July 2006–April 2007

Crop	Queries	%
Banana	111	15.6
Cassava	97	13.6
Maize	83	11.7
Groundnut	55	7.7
Tomato	54	7.6
Coffee	47	6.6
Orange	32	4.5
Sorghum	32	4.5
Rice	30	4.2
Cabbage	20	2.8
Sweet potato	17	2.4
Beans	13	1.8
Green gram	13	1.8
Watermelon	12	1.7
Cowpeas	10	1.4
Eggplant	8	1.1
Pawpaw	8	1.1
Sesame	7	1.0
Avocado	6	0.8
Passion fruit	6	0.8
Onions	5	0.7
Eucalyptus	4	0.6
Millet	4	0.6
Cereals	4	0.6
Citrus	4	0.6
Jackfruit	3	0.4
Soybean	3	0.4
Bitter berries	2	0.3
Cotton	2	0.3
Mango	2	0.3
Potato	2	0.3
Yams	2	0.3
Anona (Kitaferi)	1	0.1
Cocoa	1	0.1
Elephant grass	1	0.1
Finger millet	1	0.1
Green pepper	1	0.1
Kulekula nuts	1	0.1
Lemon	1	0.1
Pineapple	1	0.1
Pumpkin	1	0.1
Red pepper	1	0.1
Sugarcane	1	0.1
Vanilla	1	0.1
No information	1	0.1
Total	711	100

Table 1.5 Uganda – selected plant health problems from three clinics, July 2006–April 2007

Crop	Diagnosis	No.
Banana	Fusarium wilt	3
	Banana weevil	8
	Banana wilt	21
	Bacterial wilt	71
Cabbage	Black rot	5
Cassava	Mites	5
	Fungus	6
	Cassava brown streak virus	6
	Mealybug	7
	Root rot	18
	Cassava mosaic	43
Citrus	Tristeza	3
	Leaf spot	3
	Fungus	5
	Fruit fly	6
	Leaf miner	14
	Scab	14
Coffee	Coffee wilt	41
Groundnut	Aphids	4
	Drought	4
	Soil-borne disease	4
	Rosette	37
Maize	Nutrient deficiency	6
	Smut	7
	Striga	56
	Stalk borer	10

intrigue and confuse plant doctors at first: for many this is the first time they have received unsolicited queries. Over 100 crops have been received at the clinics in Bolivia, for example, with around 300 different problems that have abiotic as well as biotic causes (Bentley et al. 2009a).

The plant doctors get many things right but the clinic data also reveal the limits of their knowledge in detecting problems as well as selecting good advice. Reviewing recommendations is a sensitive matter since it is easy to undermine confidence of new doctors. Experiences in countries such as Bangladesh (Danielsen and Kelly 2009) show that the plant doctors are eager to learn how to improve. They are also keen to learn new skills, such as recognition of virus diseases, for example. The systematic collection and review of clinic data expose and confirm demands that other advisory methods are unlikely to reveal for such a wide variety of crops and farmers.

Research scientists with expert knowledge of a few crops can be daunted by the scope of the problems presented at the clinics. And being accountable to farmers is for many a new experience. A quote from a series of essays about the plant doctors of SOFRI in Vietnam (Kelly and Danielsen 2008) illustrates the challenge for scientists in working at a plant clinic:

> Truc has been a plant doctor for over a year, and is very happy to share her knowledge with farmers but admits that at first she didn't feel confident. With a biotechnical background, her training was mainly laboratory based, and Truc had little field experience. She sought advice from other plant doctors, read many books on field symptoms, and reviewed the clinic register to find out more. "Now I feel like a teacher as I train farmers and have many students. It is a good feeling to be a plant doctor.

In Nicaragua, Dimas Sarantes, an agricultural officer or técnico, had just obtained his first job with a farmer co-operative when he found himself involved with a plant clinic initiative (Danielsen and Fernández 2008):

> At that time I was just finishing up at the university ... I had already sent my CV to the co-operative, and the next week they called me for a job interview. By the following Friday I was receiving farmers at the plant clinic.
>
> We got training with the Global Plant Clinic and the National Commission for Agricultural Education (CNEA). Today we give a more accurate diagnosis, and we are not sending samples so often to the laboratory. We have been running the clinic for two years now, and it is bearing the fruit we all hoped for. In November 2007 we graduated as 'plant doctors', it's strange, but that's how it is. This has been one of my main achievements, an important added value to the work I do.

1.4 Plant Health Systems

Nicaragua has the most extensive plant clinic operations out of all the countries where the GPC supports clinic schemes (Table 1.1). See Danielsen and Fernández 2008 for a full description of partners and activities. By the end of 2007, 3 years after the first *Puestos para Plantas* ('plant health stalls') began, there were 18 clinics serving farming communities from San Juan del Río Coco in Las Segovias to Somotillo in El Occidente and Masaya close to the capital city Managua. The growth in numbers of clinics prompted the creation of a Plant Disease Management and Diagnostic Network, proposed and developed by Nicaraguan scientists, extension workers, university teachers and the phytosanitary authorities. The Nicaragua experience is described by Danielsen and Fernández (2008), who also reviewed achievements and performance of the clinics.

Although Bangladesh has a similar number of clinics to Nicaragua, there has been a slower integration of the individually successful clinics with other organisations that work on pests and diseases. Links to diagnostic laboratories are slowly improving as scientists understand the need to balance scientific priorities with establishing a reliable service for plant health clinics and serving farmers directly. The official government extension service in Bangladesh is keen to work more closely with the clinics but their involvement is hampered by competing demands to undertake other tasks and to respond to emergencies, such as flooding. Building relationships and creating robust links to the clinics requires patience and persistence, supported by solid evidence of clinic results and benefits delivered.

Bolivia has also made good progress in creating a plant health system, albeit one built around Ladiplantas, a community plant clinic based in Comarapa and managed by CIAT Santa Cruz (Bentley and Boa 2004). CIAT manage five clinics and have their own diagnostic laboratory as well as technical staff to visit farmers in their fields. This integration of effort has strengthened the support by the Department of Santa Cruz (who fund CIAT) to farmers, combining regular and reliable advice in the clinics, with technical backup from Ladiplantas and follow-up visits in the field.

1.5 Surveillance and New Disease Records

Before the Global Plant Clinic began its programme of establishing plant clinics (Boa 2009), CABI offered an expert diagnostic and advisory service, also supported by DFID. Scientists would send samples to the UK for identification and recommendations on how to control problems. This service has continued and is a valuable component of the support the GPC provides to the plant health clinics and country schemes. The number of new disease records has increased significantly (Table 1.6), although discovery of new diseases through farmer queries is still a relative small proportion of the total published. A total of 40 new disease records have been published to March 2009, all in the journal 'Plant Pathology'.

The clinics have a wider role in documenting current and emerging plant health threats (Table 1.5). The accuracy of diagnoses needs to be reviewed and unknown or unclear problems resolved. Examples recorded by plant doctors in clinic registers include 'fungus on cassava' or 'soil-borne disease on groundnut'. Both statements are incomplete and need resolving in order to review any recommendations given.

Table 1.6 New disease records from the GPC published in plant pathology according to IPPC region, 2001–2009 (March)

IPPC Region	Countries	NDRs
Africa	8	15
Asia	4	7
Europe	1	2
Latin America and Caribbean	7	15
Near East	1	1
North America	0	0
Total	21	40

Fig. 1.4 Farmers in Western Kenya attend an impromptu 'Going Public' session by the road. The symptoms of napier grass stunt, a damaging disease which has affected fodder production, are explained and they ask questions about how to control it

1.6 'Going Public' and Public Plant Health Campaigns

Plant health clinics have a fixed location. Their outreach varies but initial estimates are of a potential audience of 2,000 people in more densely populated areas. Clinics have regular contacts with thousands of farmers each week; nine clinics operating over a period of 6 years in Bolivia have attracted more than 9,000 queries. Yet there are many more who will not visit the clinic because they live beyond the outreach area, do not know the clinic exists, or are unable to visit at the appointed time.

'Going Public' is a mass extension method that was first used in Bolivia (Bentley et al. 2003). It has since been used in Uganda (Boa 2005) against banana xanthomonas wilt and in Kenya against napier grass stunt (Boa et al. 2005) (Fig. 1.4), a new phytoplasma disease that the GPC helped to confirm and study. In 4 days a team of four people visited 13 different sites in Western Kenya and explained how to identify napier grass stunt, how to control it and answered questions from a total audience of nearly 800 people.

'Going Public' is used by AAS in Bangladesh to extend the range of plant health services to communities that are not served by plant clinics. In Bolivia, Oscar Díaz of Proinpa used 'Going Public' regularly in market places to explain about methods for controlling major potato pests and diseases. 'Going Public' is the basis for Public Plant Health Campaigns (Jornadas de Salud de Plantas) which were developed in Nicaragua in 2008. These 'Jornadas' have been successful in their own right as well as reinvigorating the clinics (Danielsen and Colmenarez 2009)

1.7 Ethnopathology

Farmers describe plant health problems using a different vocabulary to scientists and agriculturists. The meanings of words used to describe these problems were analysed in three separate ethnopathology studies carried out in Uganda, Bangladesh and Bolivia. The results have now been published (Bentley et al. 2009b) and will help plant doctors understand better the potential causes of problems and to learn more about them.

1.8 Conclusions

Plant health clinics are responsive and accountable. After 6 years of running clinics and testing models in 21 countries important lessons have been learned on how best to organise clinics, train plant doctors to offer a reliable service to farmers and provide the basis for building plant health systems that integrate efforts and resources for increased impact on farmers' livelihoods.

Acknowledgements The GPC is managed by CABI in alliance with Rothamsted Research and the Food and Environment Research Agency, both based in the UK. The GPC is generously supported by the UK Department for International Development. I wish to thank my principal colleagues for their important contributions to country schemes and strategic development of the GPC: Jeffery Bentley, Solveig Danielsen, Rob Reeder, Paula Kelly, Rob Harling, Phil Jones and Yaima Arocha. Many others have also contributed to the running of the plant clinics in-country and several appear as authors in GPC references shown below.

References

Anderson JR (2007) Agricultural Advisory Services. A background paper for 'Innovating through science and technology', Chapter 7 of the World Development Report 2008. World Bank

Bentley JW (2009) Impact of IPM extension for smallholder farmers in the tropics. In: Peshin R, Dhawan AK (eds) Integrated pest management: dissemination and impact. Springer, The Netherlands, pp 333–346

Bentley JW, Boa ER, Van Mele P, Almanza J, Vasques D, Eguino S (2003) Going public: a new extension method. Int J Agric Sust 1:108–123

Bentley J, Boa E (2004) Community plant health clinic. An original concept for agriculture and farm families. Global Plant Clinic, CABI E-UK, Egham. www.research4development.info

Bentley J, Boa E, Danielsen S, Zakaria AKM (2007) Plant clinics for healthy crops. Leisa Mag 23(4):16–17

Bentley JW, Boa E, Danielsen S, Franco P, Antezana O, Villarroel B, Rodríguez H, Ferrrufino J, Franco J, Pereira R, Herbas J, Díaz O, Lino V, Villarroel J, Almendras F, Colque S (2009a) Plant Health Clinics in Bolivia 2000–2009: operations and preliminary results. *Food Security* 1:371–386

Bentley JW, Boa ER, Kelly P, Harun-Ar-Rashid M, Rahman AKM, Kabeere F, Herbas J (2009b) Ethnopathology: local knowledge of plant health problems in Bangladesh, Uganda and Bolivia. Plant Pathol 58(4):773–781

Boa E, Ajanga S, Mulaa M, Jones P (2005) Going Public in Kenya. How to entertain and inform lots of people about napier grass stunt, a new threat to dairy farmers in East Africa. Global Plant Clinic, CABI E-UK, Egham, 14 pp. www.research4development.info

Boa ER (2005) Going public: extension in markets. Rural Dev News 2(2005):42–44

Boa ER (2009) How the global plant clinic began. Outlook Pest Manag June:112–116

Campbell CL, Peterson PD, Griffith CS (1999) The formative years of plant pathology in the United States. The American Phytopathological Society, St. Paul, MN

Danielsen S, Kelly P (2009) Plant clinics uncovered. Global Plant Clinic, CABI E-UK, Egham. www.research4development.info

Danielsen S, Fernández M (2008) Public plant health services for everyone. FUNICA, Managua. Spanish at www.funica.org.ni; in English at www.research4development.info

Kelly P, Bentley J, Harun-Ar-Rashid ZAKM, Nuruzzamann M (2008) Plant clinics help curb pesticide use in Bangladesh. Pesticide News 81:5–6

Kelly P, Danielsen S (2008). Fruitful Plant Clinics. Stories about plant doctors and patients from Viet Nam. Global Plant Clinic, CABI E-UK, Egham. www.research4development.info

Large EC (1940) The Advance of the Fungi. Jonathan Cape, London

Miller SA, Beed FD, Harmon CL (2009) Plant disease diagnostic capabilities and networks. *Annual Review of Phytopathology* 47:15–38

Richards P (1985) Indigenous Agricultural Revolution. Ecology and food production in West Africa. Hutchinson, London

Smith J, Waage J, Woodhall JW, Bishop SJ, Spence NJ (2008) The challenge of providing plant pest diagnostic services for Africa. Eur J Plant Pathol 121:265–275

Van de Berg H, Jiggins J (2007) Investing in farmers: the impacts of farmer field schools in relation to integrated pest management. World Dev 35:663–686

Chapter 2
Participatory Approaches and Plant Diseases in Less Developed Countries

J.G.M. Vos, S.L.J. Page, and U. Krauss

2.1 Introduction

Globalisation and increasing demand for safe food, for example free from pesticide residues, and sustainable production practices, is increasing the need for farmers across the world to acquire new knowledge continually on crop production practices. Farmers on small holdings, particularly women, in resource-poor countries and communities often do not have ready access to information specific to their requirements for crop management and protection. They generally receive information through informal sources, such as neighbours, family and friends, and the agricultural industries. In many cases, staff from commercial organisations are more regular conveyors of information than the government extension service.

Following reductionist principles, scientific research on pest management tends to focus on single pests in isolated systems, even if done using integrated pest management (IPM) concepts. Old-style extension relies on the over-simplified theory that scientists are those who the sources of new knowledge, that extension workers are those who transfer the knowledge and that farmers are those who adopt/reject new knowledge.

The Green Revolution in Asia in the early 1970s came about when higher yielding cereal cultivars were introduced as a major problem-solving technology. A dramatic success in terms of increased yields resulted in reduced food shortages, but the newer cultivars needed increased inputs such as water, pesticides and fertilisers.

J.G.M. Vos (✉)
CAB International (CABI), Kastanjelaan 5, 3833, AN Leusden, the Netherlands
e-mail: j.vos@cabi.org

S.L.J. Page
CAB International (CABI), Bakeham Lane, Egham, Surrey TW20 9TY, UK
e-mail: s.page@cabi.org

U. Krauss
CAB International (CABI), Gordon Street, Curepe, Trinidad and Tobago
e-mail: u.krauss@cabi.org

N.V. Hardwick and M.L. Gullino (eds.), *Knowledge and Technology Transfer for Plant Pathology*, Plant Pathology in the 21st Century 4,
DOI 10.1007/978-1-4020-8934-3_2,

The use of pesticides caused risks to farmers, farm workers as well as consumers and the centralised approach (including state-managed procurement of inputs) became increasingly criticised, leading to a demand for a new paradigm of feeding the world through *sustainable* food production (Schillhorn van Veen 2003).

Nowadays, farmers have to deal with increasingly competitive markets that demand high quality produce at a low price. At the same time more pressing production problems such as decline in soil fertility and water shortages are emerging in many areas of the world. Contributing to the problem, pests (including diseases and weeds) are adapting to break through (low-cost) single technology-style crop protection. Holistic approaches to knowledge generation and dissemination are therefore required.

The agricultural scientific community has an important role and duty to contribute to knowledge transfer and rural development. Stiglitz (2007) states:

> we recognize that knowledge is not only a public good, but a *global* or *international* public good. We have also come to recognize that knowledge is central to successful development. The international community … has a collective responsibility for the creation and dissemination of one global public good – knowledge for development.

Today, extension workers have to collaborate with multiple sectors, each with their personal and institutional histories, norms, values and interests (after Van Mele et al. 2005a).

2.2 Complex Messages

The transfer of pest management technologies is complex, because it needs to contribute to an overall sustainable solution to a crop production problem and fit into a sustainable production system.

There is an ongoing debate about what can actually be classified as 'sustainable'. Marketing schemes, such as organic certifications, would represent ecologically sound practices, but there is less emphasis on the cultural and/or social equity. In the system of fair trade, those same cultural and social aspects are well managed, but these schemes may pay less attention to the stability of the agro-ecosystem. However, in general there is agreement that the implementation of sustainable agricultural practices should contribute to:

- Economic development
- Food security
- Human development/people empowerment
- Stable environment

… and be characterised as:

- Ecologically sound
- Based on a systems approach (location specific and continuously evolving)
- Economically viable
- Culturally appropriate
- Socially just and equitable

It is evident that single-message transfer style extension will not work for complex subjects such as sustainable agricultural production. One of the recognised shortcomings of traditional, top-down methods of extension, as practiced during the days of the green revolution, has been that blanket recommendations were conveyed without sufficient reference to local conditions (both agro-ecologically and socio-economically), or to local knowledge.

Agricultural knowledge generation and transfer is a complex interaction of various elements (Fig. 2.1)

To achieve what is called 'Putting Knowledge to Work' (Holderness 2003), knowledge established externally through formal science and that derived locally through a community's own experiences are necessary for successful innovation. It is argued that both types of knowledge have value, with scientific knowledge obtained and trusted through scientific process or other external validation, while indigenous knowledge is validated and trusted by the experiences of the community itself.

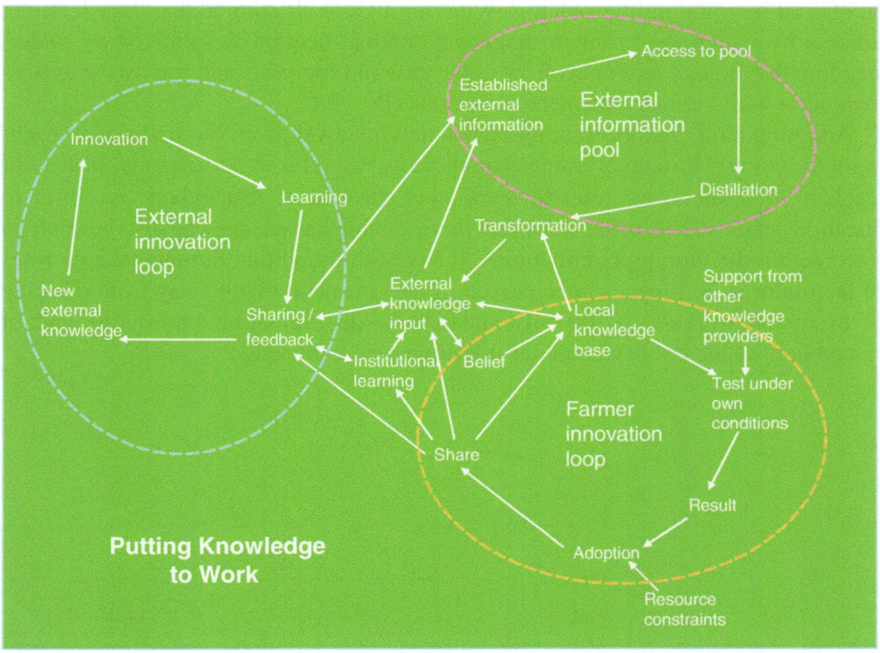

Fig. 2.1 The process of knowledge and information generation and transfer (After Holderness 2003). This diagram pictures three sources of knowledge and information and how these should interlink for appropriate distribution of knowledge. The External innovation loop is generated through off-farm research by specialised scientists in the public and private domain. The External information pool consists of documentation (incl grey material) of research findings and others, generally stored in libraries at institutions and universities. The Farmer innovation loop is generated through on-farm research activities by farmers

2.3 Participatory Approaches

Participatory methods are currently widely used and there is a variety (and confusion) of terms and acronyms to name numerous different applications. The common denominators amongst all are the focus on active participation of the target/end-user groups in discovering new knowledge and the focus on the facilitating role of trainers. Hence, participatory approaches are learner-centred and enhance ownership of findings, thereby improving the understanding of underlying principles of, for example pest management. It is important to distinguish the situations appropriate for participatory training and those where participatory research would be more appropriate (Fig. 2.2)

2.3.1 Participatory Training

Participatory training is applicable when knowledge is available, either at the farmer level (indigenous knowledge) or at the research level. Such knowledge can be offered by training in problem solving through facilitating a discovery learning process by which the problem diagnosis or identification is followed by understanding the biology and/or ecology in the case of crop pests and finally the experimentation with various management options (Box 2.1).

Adopting participatory training means moving away from the traditional teaching terminology. Instead of trainees or students, we refer to participants. Instead of teachers or trainers, we refer to facilitators. Instead of researchers, we refer to resource persons. This reflects the fact that valuable knowledge can also be sourced from within the farming communities. It has been found that farmers take on roles of facilitator as well as resource person. It also means moving away from thinking of extension workers as the only stakeholder disseminating knowledge and of researchers as the only stakeholder inventing new knowledge.

		What Farmers	
		Know	Don't Know
What Scientists	Know	Common Knowledge	Participatory Training
	Don't Know	Indigenous Knowledge	Participatory Research

Fig. 2.2 'Johari window' representing knowledge between scientists and farmers

Box 2.1 Managing bacterial wilt in tomato, Vietnam

Farmers used to recognise wilting tomato plants in their fields, but did not practice roguing as they lacked the understanding that wilting plants in the field become sources of infection for other plants.

The classic exercise of cutting wilting plants at the stem base and inserting a piece of stem in a glass of water to see the bacterial white milky ooze come out was practised with farmers in a field training session.

(continued)

Box 2.1 (continued)

Farmers were excited at the discovery but queried whether this was the disease that was killing their plants. A follow-up experiment was done using two recently potted healthy young plants. The glass with the water and bacterial ooze was emptied on the soil of one potted plant and a glass with clean tap water was emptied on the soil of the other potted plant. The two plants were monitored and after 9 days the results were clearly visible: the infected plant showed the wilt whereas the control plant remained healthy.

This simple exercise opened the minds of the farmers. First and foremost, farmers started to understand the spread of the disease in water as well as the fact that wilted plants were sources of infection. Facilitators steered the discussion so that farmers came to the conclusion by themselves that such wilting plants should be removed from the field as soon as symptoms became apparent.

The *training process* is the methodology used in knowledge transfer, based on adult learning principles:

- Adults are voluntary learners
- Adults are motivated to learn only when the subject is of interest to them
- Adults exchange good experiences
- Adults learn best when actively involved and discover by themselves
- Adults need a real-world approach

There are social dimensions to group versus individual training. The impact of the farmer-field school process is specifically high in cultures where farmers like to work together. In other situations, farmer exchange visits or other forms of group meetings appear to have more appeal, depending on costs.

The *training content* is the actual message that needs transferring. In the 'spray dye' exercise (Box 2.2) the message is that one can save on pesticide use while increasing efficacy with adaptations to the spraying equipment.

It is our experience that participatory training leads to a high level of enthusiasm with participants and the desire to learn more. It is also our experience that one-off training does not necessarily build sufficient confidence for farmers to continue with the discovery learning process beyond the training programme. Following up with related activities is more likely to encourage continued innovation and help farmers see new opportunities for change (Box 2.3).

The follow-up can range from establishing a network of trained farmers, organising regular meetings to stimulate continued information exchange. Marketing activities, literacy classes and beekeeping are all examples of further activities seen in the field. One particularly interesting development is where farmers start doing research in their own fields on unresolved problems. Facilitators and resource persons should continue to be key here in supporting the process of selecting comparable treatments as well as the process of trial design, trial monitoring and trial evaluation.

Box 2.2 Rational pesticide use in a perennial crop

In Cameroon, smallholder cocoa farmers regularly apply fungicides to control cocoa black pod, caused by *Phytophthora*. When farmers use pesticides in cocoa, the tendency is to spray until run-off, which is inefficient and ineffective. In addition, because of the tropical conditions, they don't use protective clothing to avoid contamination with the spray and poisoning.

(continued)

Box 2.2 (continued)

A spray dye exercise was done: farmers were wrapped in white paper (tissue paper or flipcharts) and their sprayers (preferably of a variety of type and age) filled with water coloured with red food dye. Some farmers were asked to spray by their usual method and some were asked to spot apply only as if they were spraying to control pod rot. After 10 min, all farmers who had used the sprayers were gathered together and the contamination by the red dye on their white outfits were studied and discussed.

Old sprayers generally leaked and the red dye on the sprayers' hands was apparent when they adjusted their nozzles while spraying. In addition, they also measured the amount of 'pesticide' that had been used over the 10 min exercise. Farmers then realised that they were at risk when using old and leaking knapsack sprayers. They also discovered that they could save on costly pesticides through targeted applications and by using well-maintained spray equipment.

Box 2.3 Cotton grading in Zimbabwe

Organic cotton farmers in Zimbabwe had participated in farmer field schools in order to learn about non-chemical pest control and the need to grade their harvest before transporting it to market. The farmers were only convinced of the need to grade the cotton once they had visited the local cotton gin (industry to separate cotton from its seeds) and spoken to the manager (Page 2000).

2.3.2 Participatory Research

Participatory research is applicable in areas where there are no known solutions for farmers' problems or when scientific recommendations conflict with traditional practises. Many definitions are available, but generally farmers set the agenda, evaluate and develop technologies under their own conditions with assistance from facilitators and resource persons.

The process in participatory research doesn't differ much from participatory training. However, the outcome is uncertain and farmers need to be leading and/or play a more proactive role in the field activities. Particularly important in participatory research is the trial design, choice of parameters to observe as well as the evaluation. It is suggested therefore that these activities become joint activities between resource persons and farmers. Resource persons need to provide support to make sure that, for example a proper control is included as one of the treatments (even scientists sometimes forget to do so!) and that the trial design allows for proper analysis on completion.

Careful consideration should be given to the number and kind of parameters that are chosen. Farmers may wish to observe only those parameters that they have learnt to observe before, for example those they were trained to monitor during a farmer-field school, such as number of infected plants. Resource persons on the other hand may wish to observe many more and different parameters such as infection severity. A compromise should be reached, with practical and 'hands-on' observations that are likely to answer the farmers' research question at the end of the trial.

The observations should be done by farmers, but resource persons will want to assist with the analysis. A good compromise is for the resource persons to analyse the data using appropriate statistics and then repeating the exercise with farmers following a simplified statistics method. Subsequently, farmers need to be asked whether they agree with the outcome and why they think the outcome is as such. This will lead to a good debate and to further understanding of the problem as well as potential ideas for follow-up research, by farmers, resource persons or both (Boxes 2.4 and 2.5).

Box 2.4 Discovering about vegetable nematode management in Ghana

Vegetable production in Ghana is suffering from root-knot nematodes in many areas. Discussions with farmers and extension workers showed that local knowledge included the use of chicken manure to reduce root-knot nematode problems.

To verify this and clarify that properly matured compost is free from plant-parasitic nematodes, a field study was done with treatments of organic (chicken) manure and inorganic fertilisers compared to planting in compost and in soil without fertiliser (control). Trainees monitored tomato growth and

(continued)

Box 2.4 (continued)

production and found that the treatments with inorganic fertilisers and the control didn't grow well and hardly produced. The treatment with chicken manure produced well, as did the treatment where tomato seedlings were planted in compost.

The real discovery came, however after the end of the season, when the plants were uprooted and trainees saw the difference in the root systems between the different treatments. They discovered that where tomato roots grew in compost, they were healthy, thus giving the plants a good start. They also found that in the chicken manure treated plot the roots were infected but not as badly as in the inorganic fertilised or control plot.

This discovery led to a change in thinking about soils and crop nutrition in relation to crop health.

Box 2.5 Integrating old and new ideas to manage frosty pod rot on cocoa in Central America

In Costa Rica, cocoa growers loose 80% of their cocoa pods to frosty pod rot (caused by *Moniliophthora roreri*). Traditionally, the disease is managed by phytosanitation, that is the regular removal of infected pods. Researchers have repeatedly recommended increasing the phytosanitation frequency from monthly to weekly intervals. However, this recommendation contradicted the growers' intuition, who did not see this labour investment as being economically viable.

Box 2.5 (continued)

In a participatory evaluation of cultural and biological control in organic smallholdings, weekly and fortnightly phytosanitation and seven biological treatments, using a range of *Clonostachys* and *Trichoderma* species, were tested. Weekly phytosanitation reduced frosty pod rot significantly, probably via a reduction in sporulation of the fungal pathogen. Both regimes increased yields but only weekly phytosanitation augmented the percentage of healthy pods. Four of the biological treatments (including combinations of the biocontrol species) reduced moniliasis, with yield improvements of up to 50%.

Farmers opted to combine cultural and biological control for best results. Even then, weekly phytosanitation proved necessary, that is biocontrol provided no single solution. But with effective biocontrol, pod sporulation is inhibited and the phytosanitary operation itself becomes gradually less labour-intensive. Farmers needed to experience this for themselves in their own crops, in order to counter their instinct.

The experiential learning circle of Kolb (1984) is relevant here, which should lead to continuous problem-solving, in which (1) immediate or concrete experiences provide a basis for (2) observations and reflections. These 'observations and reflections' are assimilated and distilled into (3) abstract concepts producing new implications for action that are (4) actively tested in turn, creating new experiences (Fig. 2.3).

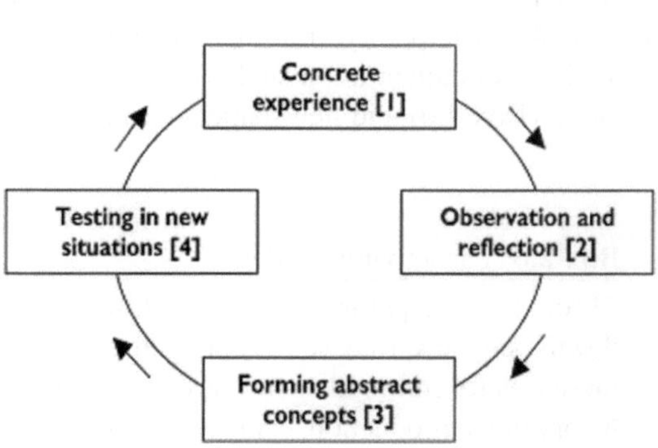

Fig. 2.3 The experimental learning circle (Kolb 1984)

2.4 Regarding Impact

Impact assessment can also be a participatory exercise with farmers and scientists. In order to help this process, farmers should be encouraged to keep their own records. Key indicators of success can initially be selected by farmers during group discussions and facilitators should then prepare short (1–2 page) forms that can be filled in using numbers and a few simple words. Scientists may prepare outlines. However, farmers and facilitators need to 'translate' these into their own forms. Completed forms can be photocopied so that both farmers and researchers can use the data to monitor progress and assess impact. In addition to this, the recording of planting and harvesting dates, pest out-breaks, input, transport and marketing costs, will enable farmers to calculate whether or not they are profiting from their chosen input strategy. Once farmers realise the value of record-keeping they will be eager to continue the process in their own exercise books.

FPR has been criticised for its limited scale of coverage (Farrington 1998). This is because participatory processes work well on pilot scales but time and again human and financial resources prove to be too limited for scaling-up. Conroy and Sutherland (2004) suggest using the 'recommendation domain' which defines the target population as a tool for developing a strategy for scaling-up the benefits from farmer-participatory activities. The size of the recommendation domain will depend on the following:

1. How widespread is the production constraint or opportunity.
2. The number of households involved in producing the relevant commodity or with a similar problem.
3. The resources (land, labour and money) available to the household producing the commodity.
4. The likely availability of the inputs needed.

Project reports, survey and census data can provide most of the information needed to determine the number of households to include in the domain.

Searching for cost-effective methods to ensure impact beyond pilot sites have led to some good examples of innovative approaches to disseminating successful results of participatory training/research. These include using farmer networks, excursions, documentation such as fact-sheets and manuals, and use of media, such as video (Box 2.6) and newspapers (Box 2.7).

Box 2.6 Participatory video

Through participatory research with scientists, women farmers in Bangladesh developed new rice seed management methods that gave consistent yield increases for minimal cost. They then shared their knowledge through participatory video production, using local languages and practical illustrations directly relevant to other women. These methods enabled messages to quickly reach thousands of women (Van Mele et al. 2005b). The programme was awarded an International Visual Communications Award for its innovative approach.

Impact assessment amongst 115 women who had watched the video twice or more at village venues revealed that these women had been able to implement the new methods and thus gained an average increase of 20 additional food secure days at no extra cost (Page et al. 2008).

Box 2.6 (continued)

Currently six videos on rice seed and rice post-harvest management, have already reached over 500,000 farmers in four Asian and 10 African countries, and to over 40 million people via television in Bangladesh and The Gambia (Van Mele P, personal communication, Africa Rice Center).

Box 2.7 Cocoa farmers' newspaper in Ghana

The Ghanaian Cocoa Farmers Newspaper was developed to transfer knowledge to smallholder cocoa farmers all over Ghana, informing them about good agricultural practice in cocoa. For this bi-annual newspaper, articles are produced and edited in close collaboration with the Cocoa Research Institute Ghana, illustrated by an artist and printed by the Daily Graphic in Accra. The first edition came out in 2006, with 70,000 tabloid copies distributed through licensed cocoa buying companies who had agreed to assist with distribution to the farmers. Initial surveys have taken place to assess the impact and relevance to farmers and regular publication of such a newspaper is expected to be an effective communication tool, contributing to rural development and the future of Ghana's cocoa industry (Keith A Holmes, personal communication, CABI).

Notwithstanding these successful larger scale examples, more attention is still needed for documentation of processes and cost-benefit analysis as well as impact assessment. Lack of such impact studies and documentation is probably due to the limitation in funding sources, both in terms of time (project duration not sufficient to assess post-project impact) and scale (cost of measuring impact on knowledge dissemination to tens of thousands of farmers and more).

2.5 Discussion

The latest development paradigm places emphasis on local knowledge for development. It gives priority to partnerships and emphasises participation, not just by government agencies, but also by non-governmental organisations, and other parts of civil society, as the best way to achieve sustainable development. Adopting participatory design approaches means defying prescriptive methods and techniques, which makes successful implementation a complex challenge (Scarf and Hutchinson 2003). Due to the complexity of community dynamics as a human process there are no blueprints, nor ready made recipes of participatory processes that can be applied to promote participatory development (Botes and van Rensburg 2000).

Duraiappah et al. (2005) argue that:

> there is no doubt that the introduction of participatory approaches to development over the past three decades has effectively demonstrated the capacity of men and women from poor communities to participate actively in research, project design and policy analysis.... As in all research processes, the potential for researcher bias exists. Due to the power imbalances inherent in participatory development, and the often sensitive and critical nature of the issues being addressed through participatory research, care and attention must be taken to ensure that these processes provide benefits and enhance the capabilities and freedoms of the poor.

As advised by Bentley and Baker (2002), if the participatory approach is to become part of mainstream research, then it should be taught in universities and also in agricultural colleges.

Looking at improving extension, the Neuchatel Group (2006) argues for:

- Services to be driven by user demand
- Service providers to be accountable to the users
- Users to have a free choice of service providers

They conclude that preconditions for success are enabling policies and public sector commitment to the transition but also that the public sector must stop the free supply of extension services that can be delivered through the private sector (Neuchatel Group 2006).

These considerations are not confined to the developing world. A recent study in Europe concluded that sustainable rural development could be improved by paying more attention to the interaction between different types of knowledge, such as local, scientific and political (Bruckmeier and Tovey 2008). Their study shows four main ways of managing different types of knowledge in rural development in Europe:

- Resource renewal, for example, sustainable forest management, which uses scientific knowledge as a guide.
- Quality of life, for example, improving access to utilities, welfare, or aesthetics, which uses managerial or political knowledge (such as from planners).
- Improving local sustainable livelihoods, where local knowledge is used.
- Participatory resource management, whereby all stakeholders with an interest in the resource participate and no single form of knowledge dominates.

We conclude, therefore, that science has played and should play an important role in sustainable rural development and that this should continue using selected scientific knowledge in participatory processes. Key stakeholders in the knowledge transfer system need to change roles and attitudes and this includes both the public and private sector. This requires institutional change, which is being piloted across the world and described in some of the project examples that are highlighted in this chapter.

References

Bentley JW, Baker PS (2002) Manual for collaborative research with smallholder coffee farmers. The Commodity Press, Egham, UK

Botes L, van Rensburg D (2000) Community participation in development: nine plagues and twelve commandments. Commun Dev J 35(1):41–58.

Bruckmeier K, Tovey H (2008) Knowledge in sustainable rural development: from forms of knowledge to knowledge processes. Sociologia Ruralis 48:313–329

Conroy C, Sutherland A (2004) Participatory technology development with resource-poor farmers: maximising impact through the use of recommendation domains. ODI AgREN Network Paper no. 133

Duraiappah AK, Roddy P, Parry J-E (2005) Have participatory approaches increased capabilities? International Institute for Sustainable Development (IISD), Colorado. http://www.iisd.org/pdf/2005/economics_participatory_approaches.pdf

Farrington J (1998) Organisational roles in farmer participatory research and extension: lessons from the last decade. Overseas Development Institute, London. Natural Resource Perspectives, No. 27

Holderness M (2003) Putting knowledge to work: rural knowledge partnerships, a catalyst for development. Proceedings, 2nd International Conference, Global Forum for Agricultural Research, Dakar Senegal

Kolb DA (1984) Experiential learning. Prentice Hall, Englewood Cliffs, NJ

Mele P Van, Salahuddin A, Magor NP (eds) (2005a) People and pro-poor innovation systems. In: Innovations in rural extension: case studies from Bangladesh. IRRI, CABI, Wallingford, UK, pp 257–286

Mele P Van, Zakaria AKM, Bentley JW (2005b) Watch and learn. Video education for appropriate technology. In: Van Mele P, Salahuddin A, Magor NP (eds) Innovations in rural extension: case studies from Bangladesh. IRRI, CABI, Wallingford, UK, pp 77–88

Neuchatel Group (2006) Demand Driven Agricultural Advisory Systems. http://www.neuchatelinitiative.net/english/documents/DemandDrivenAgriculturalAdvisoryServices.pdf

Page SLJ (2000) Zambezi valley organic cotton project. In: Stoll G (ed) Natural crop protection. CTA, Wageningen, The Netherlands, pp 305–313

Page SLJ, Ar-Rashid H, Zakaria AM, Dodsworth E (2008) The good seed initiative: improving food security for the poorest households in Bangladesh through the use of 'Women-to-Women' videos. Presented at the World Conference on Agricultural Information and IT, Tokyo, 24–27 August 2008

Scarf C, Hutchinson K (2003) Knowledge networks for development: a participatory design approach. International Conference on the convergence of knowledge, culture, language and information technologies, Alexandria, Egypt. http://www.cfilt.iitb.ac.in/convergence03/all%20data/paper%20032-36.pdf

Schillhorn van Veen TW (2003) The World Bank and pest management. In: Maredia KM, Dakouo D, Mota-Sanchez D (eds) Integrated pest management in the global arena. CABI, Wallingford, UK, pp 435–440

Stiglitz JE (2007) Knowledge as a global public good. Worldbank. http://www.worldbank.org/knowledge/chiefecon/articles/undpk2/index.htm

Chapter 3
Technology Adoption: Classroom in the Cocoa Block

D.I. Guest, R. Daniel, Y. Namaliu, and J.K. Konam

3.1 Introduction

Seventy two per cent of the 5.5 million people in Papua New Guinea (PNG) obtain their livelihoods from the agriculture sector (Anon 2006). The sustainable generation of income by smallholders is central to food security, poverty alleviation and access to education and health services. Sustainable production also reduces pressure on environmental and land resources. The major constraint to improved smallholder outcomes in cocoa is the poor adoption of new technologies.

3.2 The Cocoa Industry in PNG

Cocoa is PNG's third most important agricultural export crop, after coffee and palm oil, contributing up to 17% of the national agricultural revenue (estimated at K168 million in 2003; 42,000 t @ K4,000/Mt.) (Anon 2006; Simatab 2007). Papua New Guinea supplies 2% of the world's cocoa, based on 2003 export figures, but 9% of

D.I. Guest and R. Daniel
Faculty of Agriculture, Food and Natural Resources,
The University of Sydney, Sydney 2006, Australia
e-mail: d.guest@usyd.edu.au; r.daniel@usyd.edu.au

Y. Namaliu
PNG Cocoa and Coconut Institute, Stewart Research Station,
P.O. Box 642, Madang, Papua New Guinea
e-mail: yaknamaliu@yahoo.com.au

J.K. Konam (✉)
Plant Health Team, Land Resources Division,
South Pacific Commission, Private Mail Bag, SUVA, Fiji Islands
e-mail: JohnK@spc.int

N.V. Hardwick and M.L. Gullino (eds.), *Knowledge and Technology Transfer for Plant Pathology*, Plant Pathology in the 21st Century 4,
DOI 10.1007/978-1-4020-8934-3_3,

the world's fine flavour cocoa, which commands a premium price. Over 80% of PNG's cocoa is produced by 150,000 families, each farming less than 5 ha of land and producing mean yields of 300–400 kg/ha of dried beans per annum (Curry et al. 2007).

Smallholder cocoa plantings increased after 1965 and by the mid-1980s contributed approximately 70% of national production. A fall in the world cocoa price in the late 1980s saw the decline in the plantation sector and a reduction in inputs, such as fertiliser and fungicides (Connell 1997). Smallholder production is characterised by low yields, reflecting inadequate management inputs and high losses due to disease, particularly Phytophthora pod rot and canker (*P. palmivora*), vascular-streak dieback (VSD, *Oncobasidium theobromae*) and pink disease (*Erythricium salmonicolor* syn. *Corticium salmonicolor*) (Guest 2007; Keane 1981; McMahon and Purwantara 2004). PNG smallholder producers often regard cocoa as a means of securing 'ownership' of land and as a 'bank' that provides a source of cash to meet commitments such as school fees. Typically, smallholders grow cocoa as a source of supplementary income, and invest very little time or money into farm maintenance and long-term management, resulting in low yields and a very low cost of production that somewhat insulates them against market price fluctuations (K447/Mt dry beans in 1999; Omuru 2003). This is also due, in part, to labour availability issues and the lack of information about, and poor adoption of, new technologies (Ghodake et al. 1995; Lummani and Nailina 2001 cited in Curry et al. 2007).

Most current farm management recommendations were developed for well-managed plantations, which, because of the high cost of plantation production (K2,150/Mt dry beans, 4 year average 1995–1998; Curry et al. 2007), no longer dominate production (Connell 1997). When the plantation sector predominated in PNG, control of Phytophthora and other diseases was based on high input cultural practices, fungicides and disease resistant clones (e.g. for Vascular-Streak Dieback, VSD) (Prior 1984). Many former plantations have been returned to traditional landowners operating under different circumstances (Omuru et al. 2001). While plantations operated under more intensive input conditions with different labour sources, smallholder growers rely largely on family labour, apply minimal chemical inputs and are likely to spend less money on inputs when the cocoa price falls (Curry et al. 2007; Prior 1984).

Recognising the significant changes (or deterioration) in management practices that have accompanied the move to smallholder production, the PNG Cocoa and Coconut Institute (CCI) has focussed on the development of germplasm that is productive in low-input systems and the development of low to medium input disease management practices (Efron et al. 2005). In assessing grower practice Omuru et al. (2001) found that many farmers currently make few or no interventions to control disease, and that productivity improvement would be boosted by fostering farmer adoption of low to medium cost control options along with uptake of cultivars with resistance to disease. Disease losses are estimated at 40%, largely due to Phytophthora and VSD (Holderness 1992; Saul 1989). Reducing disease losses represents a key option for farming families to improve the productivity of existing cocoa plantings and to encourage greater investment in the crop to ensure sustainability of higher yields.

In July 2003, a National Cocoa Summit of cocoa industry stakeholders in PNG set a goal of 100,000 t of dry bean production by 2012. While this goal is ambitious from a market demand perspective, it is feasible because the world market for cocoa tends to be undersupplied. While the 2003 cocoa summit considered improvement in disease control as a critical priority for the industry, the PNG National Extension Summit in May 2004 highlighted more broadly the deficiencies in information dissemination and agricultural extension that has hampered delivery of improved technologies. The summit identified farmer-focussed approaches in the delivery of disease management improvements as a clear priority.

The adoption of technology aimed at improving the sustainability of food, fibre and fuel production is one of the greatest impediments to improving the quality of life particularly for smallholder farmers in developing countries. This paper describes and discusses a participatory research approach undertaken with smallholder cocoa farmers in PNG aimed at providing different levels of farm management tailored to be implemented by farmers according to their particular circumstances.

3.3 Smallholder Farmers in PNG

One of the main reasons that cocoa production has not increased is that farmers invest very little time or money in managing their cocoa farms and average yields have been stagnant for decades (Fig. 3.1). In a survey of cocoa farmers in East New Britain Province (ENBP) Curry et al. (2007) found that farmers changed the way they managed their cocoa depending on the age of the planting, so much so that the age of the cocoa trees can be predicted based on how frequently the farmer visits those trees.

Briefly, when the seedlings are first planted, the farmer visits the cocoa block regularly to ensure that the trees are established properly. The farmer applies weed and shade management until the trees are established. This is often because food crops are planted in newly established cocoa blocks and weeding is carried out as part of the garden maintenance. Many growers also recognised that young cocoa trees were vulnerable to overshading and to being overgrown by weeds (Curry et al. 2007). Once the trees start to flower and yield, the visits made by the farmer are limited almost only to the harvest period. Pest and disease management and inputs of fertiliser, pruning and sanitation are minimal.

One of the consequences of this pattern of management is that not only do the trees bear few pods but those pods are also heavily infected by diseases. Over 82% of farmers identified Phytophthora pod rot and canker and VSD as the most important constraints to cocoa production (Omuru et al. 2001). Curry et al. (2007) found that up to 74% of trees were affected by canker in ENBP, yet over 95% of farmers had no knowledge of pest and disease management. Low yields reflect the low

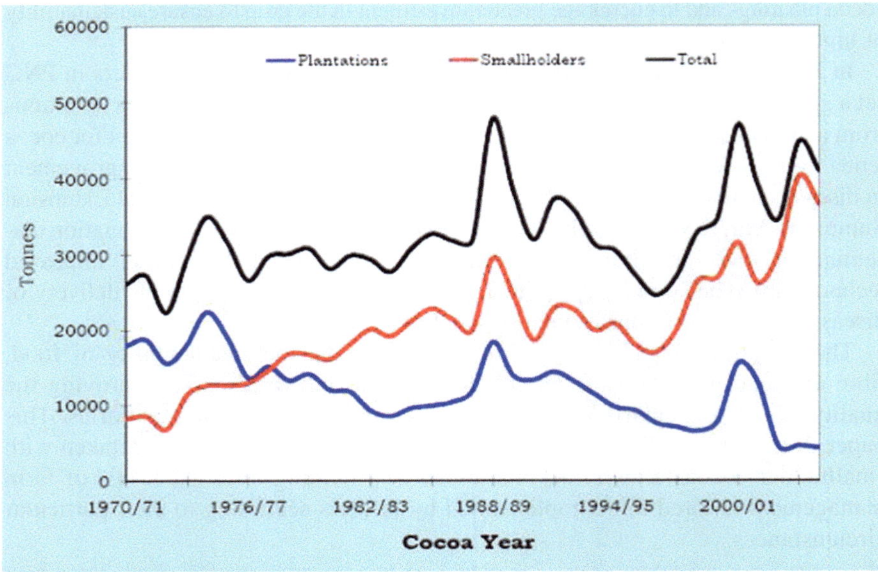

Fig. 3.1 Cocoa production in PNG by sector from 1970 to 2000/01. National cocoa production in PNG has remained at around 42,000 t dry beans per year for the last 20–30 years despite intensive research into the improvement of cocoa genotypes and cocoa production at the Cocoa and Coconut Institute of PNG (Anon 2002). This static production is likely to be due to shortcomings in the transfer and uptake of information through the agricultural extension sector, restricting the information available to farmers

management inputs, the lack of understanding of the pests and diseases, and the absence of advice on management strategies.

3.3.1 Constraints Faced by Smallholders

Smallholder cocoa farmers face very rapid social, environmental and economic changes. For many farmers cocoa is not the most important part of their lives and the daily priorities include spending time with family and friends, gardening to produce food and fishing. Cocoa, for most of the farmers, is simply a source of money when cash is needed.

The major factors that constrained production, as identified by farmers in ENBP, were the theft of cocoa pods, poor block conditions, labour shortages, shortage of farm tools and limited knowledge of correct management practices (Curry et al. 2007). Other factors included the lack of resources (e.g. pesticides or fertilisers may simply not be available, either due to lack of infrastructure, or the farmer may not have the funds to purchase them) and the farmer's past experiences (i.e. traditional methods of growing cocoa have been adequate to meet farmer's needs in the past

[Curry et al. 2007]). These factors are confounded by the limited skills of farmers and the lack of an effective extension service, which hinders the availability and transfer of new information and technology. However, when they are presented with new information, smallholder farmers are often very willing to experiment.

3.4 Participatory Action Research

National cocoa production in PNG has remained at around 42,000 t dry beans per year for the last 20–30 years (Fig. 3.1), despite intensive research into the improvement of cocoa genotypes and cocoa production at the CCI. Deficiencies in the dissemination of information and in agricultural extension limited the information that was available to farmers in PNG, and a new approach for delivery of information was required.

Participatory action research (PAR) engages farmers as both researchers and participants in the research. The aim of PAR was to identify problematic situations or issues considered by participants to be worthy of investigation in order to bring about informed changes in practice (Burns 1999, cited in Cornwell 1999).

The aim of the programme was to distribute effective disease control options to farmers. The advantages of PAR are that it offers a range of practical and cost effective options. The key is that the farmer makes the choice about which practice is most suited to his or her situation. The farmer is the person who makes the decision to adopt the particular recommendation or the options, and he or she participates in the implementation of those options. Field schools, or extension activities, were organised around a farmer's block of cocoa trees in order to engage a larger number of farmers (Fig. 3.2). In essence, the farmer becomes the researcher by choosing and testing the selected option on his or her own farm. The farmer also becomes the extension worker and disseminates the information to neighbouring farmers, all because the farmer has learnt by practical application of the chosen option (Fig. 3.3).

The first step was to gather resources and information. This was done in a number of ways, but the most effective way was to spend time with the farmers in their villages, observing their practices rather than directly interrogating them. By spending time with farmers 'true' or honest answers were derived through conversation, rather than through responses to a series of direct questions, often resulting in answers they believed the interviewer wanted to hear. This baseline information was used to identify the strengths and weaknesses of the current practices and to develop a series of new management strategies. The management options were discussed with farmers and packaged into options that gave the farmer steps to improve management at their own pace (Table 3.1). For example, the first option simply required increased labour and time inputs. It did not require that the farmer spent any money or needed to have capital available. Once the farmer engages in improved management, and has higher yields, and consequently more cash, he or she can then go to the next option, which may involve purchasing inputs.

Fig. 3.2 Cocoa management options are demonstrated in the farmer's blocks to engage a larger number of neighbouring farmers and to enable farmer participation and training in the implementation of the management inputs

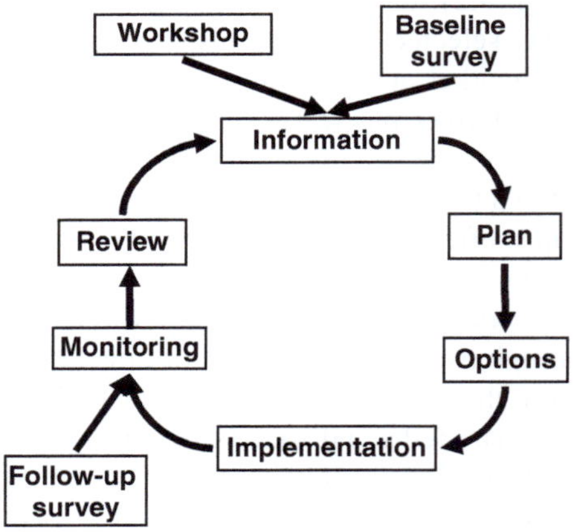

Fig. 3.3 The cycle of events in participatory action research. Workshops and baseline surveys are conducted to collect information about the current practices and level of knowledge and understanding. This enables plans and options for disease management in cocoa to be developed. Following implementation of the management options the outcomes and practices are monitored and evaluated. In this way the practices can be amended as required

Table 3.1 Integrated pest and disease management options, and corresponding expected yields based on preliminary trials at the Papua New Guinea Cocoa and Coconut Institute (CCI), developed for cocoa smallholder farmers in PNG

Input level	Activity	kg dry beans/ha
Low	• Infrequent harvests	0.7
Medium	• Cocoa and shade tree pruning • Weekly harvests and sanitation • Weed control	1.1
High	• Cocoa and shade tree pruning • Weekly harvests, sanitation & pod sleeving • Weed control • Fertilisers and manures • Canker treatment	1.8
Maximum	• Cocoa and shade tree pruning • Weekly harvests, sanitation & pod sleeving • Weed control • Fertilisers and manures • Canker and longicorn treatment • Insect vector control	2.0

Once the information was gathered from the farmers, the next stage was to develop plans for implementation of disease management strategies. New information or technology options were discussed with farmers. Constraints were identified and questions answered. The support required to implement the action plan was then identified.

Developing a series of packages of management options allowed informed farmers to improve their management gradually. For example, leaving infected pods on cocoa trees increased the inoculum level in the cocoa block sustaining high levels of disease. Implementing a strategy of harvesting not only ripe pods but also those affected by *P. palmivora*, reduced inoculum levels (Gregory and Maddison 1981). While this option was more time consuming, it required no extra cash input and was ultimately beneficial in producing higher yields of healthy pods in subsequent harvests.

3.5 Management Options for Cocoa in PNG

Recognising the need for change in management practices a series of low, medium and high-input Integrated Pest and Disease Management (IPDM) options was developed, which underpinned new farmer participation-based approaches in the delivery of new information and technologies (Konam et al. 2008). Disease losses and smallholder knowledge, skills and attitudes to disease management, fostering adoption of IPDM strategies by cocoa farmers and improving cocoa yields, were documented. This information will assist with the review and monitoring process to improve management options in the future. The goal was to transform the cocoa industry from 90% low management input to 50% medium management input, thereby increasing cocoa production and improving farmer incomes. Yields in farmer trials of up to 2,000 kg beans/ha – an eightfold increase – have been realised.

3.5.1 IPDM for Cocoa

Four options of management levels were developed after consultation with farmers (Table 3.1). The first was the traditional practice of no input; the second was medium input requiring pruning, weed management, sanitation and weekly regular harvesting; the third option added manure or fertiliser application to Option 2; while the highest level included Option 3 plus the use of herbicides for weed management, fungicides and insecticides for diseases and insect pests, including disease vector control.

Pilot trials showed that a farmer implementing Option 1 would expect around five pods per tree and the incidence of Phytophthora pod rot would be over 40% (Fig. 3.4). In implementing the second option the farmer would expect an increase in yield to around eight pods per tree, and the incidence of pod rot should decline to around 25%.

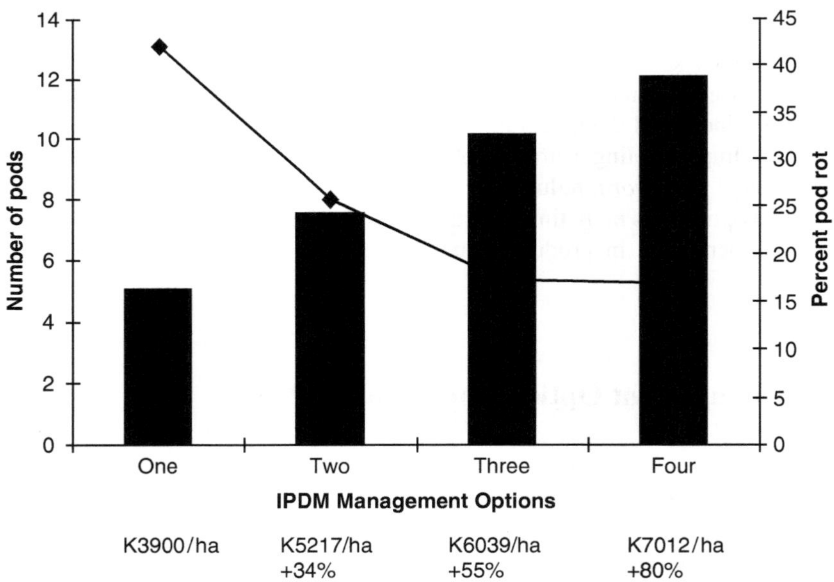

Fig. 3.4 The Integrated Pest and Disease Management options have been shown to increase yield and reduce disease incidence in cocoa in trial blocks at the Cocoa and Coconut Institute of PNG. The percentage change indicates the increase in yield compared to option 1. K/ha = value of beans in PNG Kina/ha (bars = number of pods, line = percent pod rot)

A cost-benefit analysis equated this to an increase in income of 34%. The addition of fertiliser in Option 3 increased yield to 11 pods per tree and reduced pod rod to 16%. This lead to a 55% increase in income over Option 1. In the final option the yield further increased to around 12 pods per tree and Phytophthora pod rot remained at around 16%. The benefit of Option 4 to the farmer was an 80% increase in income over the income obtained from Option 1.

3.5.2 Implementing IPDM Options Through Training Workshops and Demonstration Trials

To implement these recommendations the three main cocoa growing areas of PNG; Madang Province, East New Britain Province and the Autonomous Region of Bougainville were selected (Fig. 3.5). The cocoa management package developed contained the four options described above. In each province three sites were identified.

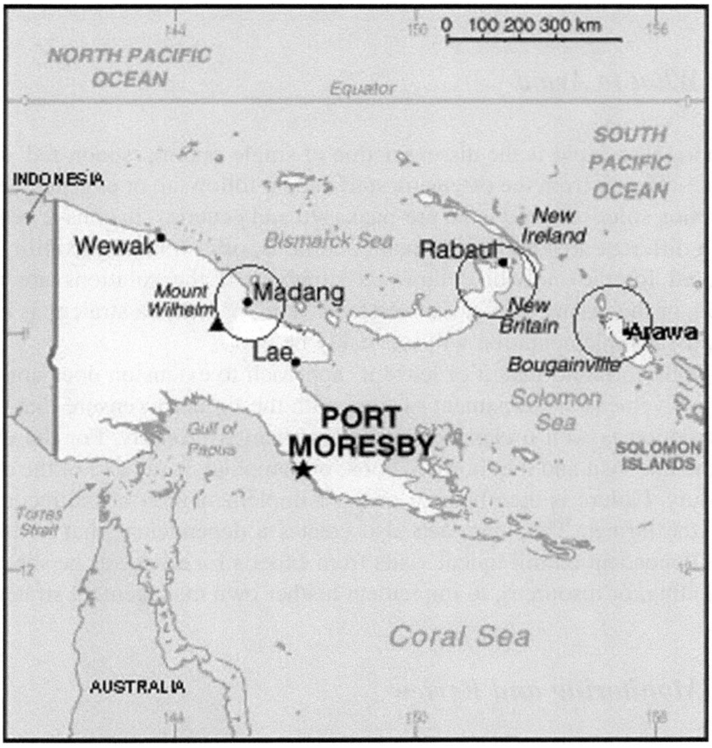

Fig. 3.5 Location of field sites in Papua New Guinea. Model farmers were selected and demonstration sites established in Madang and East New Britain provinces and the autonomous region of Bougainville (locations indicated by *circles*)

At each site 12 'model' farmers were selected (the farmers were identified during baseline surveys). The implementation phase involved a series of extension training workshops in which the extension staff and the model farmers participated in establishing the IPDM options on their farms. The community became involved as each model farmer was asked to train a further 12 farmers (known as extended farmers).

Ideally, the Option sites were established near roadsides along which many people passed each day. As the villagers passed the sites they became curious because the trees were responding to the implemented management options and they wanted to know what the farmer had done. The model farmer was engaged because he or she had participated in implementing the options and they were proud and enthusiastic in sharing their knowledge. In addition, a series of field days were run at the Option sites. The field days were followed with training for other interested farmers. The key element of this approach was that once the options were developed, the farmers decided which option was best for them and their family, they learnt the skills by training and they then taught each other. Essentially, the trained farmers became the extension agents. The farmers, by making the choice of management options they wished to implement, also owned the knowledge and were responsible for their own actions.

3.5.3 What to Avoid

It is important to avoid is the dissemination of single-option, 'spoon-fed' solutions with a one-off visit from the extension staff and no follow-up or engagement. These single-option solutions tend to be pre-packaged and generic – they have been developed in a different area under different conditions, offer limited flexibility and are not tailored for the individual farmer's situation. If the solutions are not well explained, or the farmer does not understand 'how' or 'why' a strategy is supposed to work, their implementation will inevitably be poor.

The single formula 'take it or leave it' approach to extension does not promote farmer involvement or investment of time with the farmer to ensure that the management option is well understood and implemented properly. For the extension worker the one visit and the 'how to' book, or pamphlet, is the end of the extension agent's duty. Failure is then blamed on poor implementation of the recommendations by the farmer. This approach also creates a dependency, that is the farmer becomes dependent on infrequent visits from experts for advice as he/she lacks the understanding, or resources, to implement his/her own management strategies.

3.5.4 Monitoring and Review

Essential monitoring and review of how the management options are being implemented can be done in a number of ways. Regular meetings of farmers, both those implementing the IPDM options and those continuing with traditional practice, present an opportunity to review management options and discuss what is, and what is not working, what is more difficult or easier than first expected. This discussion between the farmers is a valuable forum for review. Follow-up surveys can be conducted after 1 or 2 years to see how farmer understanding and farmer practices have changed. Baseline and follow-up studies in PNG have been conducted and are currently being analysed.

3.5.5 Case Studies

Case studies are useful because they look at how people with different backgrounds have implemented IPDM options. One farmer, Willi from the north coast of Madang became a model farmer during the initial IPDM training. He now leads a village farmer group and actively promotes investment in cocoa farming. Patrick from ENBP used to be a rascal and began cocoa farming in October 2006, and is now a model farmer. IPDM requires a commitment and so farmers have less time to get into trouble. Norman, also from ENBP, used to be a mechanic and was a good example of a cocoa farmer who only visited his cocoa to get cash. He has now found that IPDM is so successful that he has taken to cocoa farming full time.

3.6 Conclusions

The PAR approach has enabled us to identify accurately research priorities, and has provided better feedback about the problems that farmers are facing and what is and is not working. It demonstrated the impact of technology to farmers in a very direct way and gave the farmers more responsibility because they were actually involved in implementing and then explaining the improved management to their neighbours. Establishing demonstration plots and conducting field days has increased the profile of research and extension agencies, which are now much more engaged with the day-to-day problems faced by the farmers. The feedback from farmers has in turn improved the capacity of supporting researchers at CCI to focus their research on industry needs.

The question remains: 'how has this programme worked in PNG?'. In the areas where demonstration sites were established farmers report significant increases in yield, and these claims are backed up by statistics from local fermentaries. Increased yields are also reported in 'outreach' areas where the technology is spreading through farmer–farmer contact. National production reached 50,000 Mt in 2007, and the trend is upwards. It is still a long way from the target of 100,000 Mt by 2012 but the trend since the start of this project has been consistently upwards, and given the number of farmers involved, the rapidity with which the options have been adopted, and the time lag between implementation and realised benefits of improved management, it is expected that the increase in yields will continue.

References

Anon (2002) Cocoa production statistics. Cocoa Board of Papua New Guinea

Anon (2006) Compendium of food and agriculture indicators. Food and Agriculture Organisation of the United Nations Statistics Division, Rome

Burns A (1999) Collaborative action research for English language teachers. Cambridge University Press, Cambridge

Connell J (1997) Papua New Guinea: the struggle for development. Routledge, London

Cornwell S (1999) An interview with Anne Burns and Graham Crookes. Lang Teach 23:7–9

Curry G, Koczberski G, Omuru E, Nailina RS (2007) Farming or foraging? Household labour and livelihood strategies amongst smallholder cocoa growers in Papua New Guinea. Black Swan Press/Curtin University of Technology, Perth, Australia

Efron Y, Epaina P, Marfu J (2005) Breeding strategies to improve cocoa production in Papua New Guinea. In: Bekele F, End MJ, Eskes AB (eds) Proceedings of the international workshop on cocoa breeding for improved production systems, 2003. Accra, Ghana, pp 79–91

Ghodake RD, Cook KE, Kurika L, Ling G, Moxon JE, Nevenino T (1995) A rapid rural appraisal of the cocoa and coconut farming systems in the northeast lowlands of the Gazelle peninsula of East New Britain province. Department of Agriculture and Lifestock, Konedobu. Technical Report 95/1

Gregory PH, Maddison AC (1981) Epidemiology of Phytophthora on cocoa in Nigeria. CABI, Wallingford

Guest D (2007) Black pod: diverse pathogens with a global impact on cocoa yield. Phytopathology 97:1650–1653

Holderness M (1992) Biology and control of Phytophthora diseases of cocoa in Papua New Guinea. In: Keane PJ, Putter CA (eds) Cocoa pest and disease management in Southeast Asia and Australasia. Food and Agriculture Organisation of the United Nations, Rome, Italy. FAO Plant Production and Protection Paper No. 112

Keane PJ (1981) Epidemiology of vascular-streak dieback disease of cocoa in Papua New Guinea. Aust J Biol Sci 25:50–55

Konam J, Namaliu Y, Daniel R, Guest D (2008) Integrated pest and disease management for sustainable cocoa production: a training manual for farmers and extension workers. ACIAR Monograph No. 131.

Lummani J, Nailina R (2001) Tri-annual survey results for cocoa and coconut smallholders in East New Britain. PNG Cocoa and Coconut Research Institute, Kerevat and the University of New England, Armidale, NSW. Occasional Paper No 6

McMahon P, Purwantara A (2004) Phytophthora on cocoa. In: Drenth A, Guest DI (eds) Diversity and management of Phytophthora in Southeast Asia. Australian Centre for International Agricultural Research (ACIAR), Canberra, pp 104–115. Monograph No. 114

Omuru E (2003) An economic analysis of cocoa and coconut research and development in Papua New Guinea. Ph.D. thesis, Faculty of Natural and Agricultural Sciences, The University of Western Australia, Australia

Omuru E, Nailina R, Fleming E (2001) A socio-economic baseline survey of cocoa and copra smallholders in East New Britain. PNG Cocoa and Coconut Research Institute, Keravat and the University of New England, Armidale. Occasional Paper 1

Prior C (1984) Approaches to the control of diseases of cocoa in Papua New Guinea. J Plant Prot Trop 1:39–46

Saul JY (1989) A study of the resistance of Kerevat cocoa clones to pod rot caused by Phytophthora palmivora. MSc Qualifying thesis, LaTrobe University, LaTrobe, Australia

Simatab J (2007) Towards a sustainable cocoa economy in PNG: enhancing cocoa production through adoption of Integrated Pest and Disease Management (IPDM) with farmers participation. Round Table Conference on A Sustainable World Cocoa Economy, Accra, Ghana, 3–6 October 2007

Chapter 4
Scholarship of Teaching and Learning (SoTL) Projects in Plant Pathology

D.M. Eastburn and C.J. D'Arcy

4.1 Introduction

Those of us who advocate the scholarship of teaching and learning do so in order to have efforts related to teaching valued on a par with efforts related to scholarly work (i.e. research) in our disciplines. When a researcher has a 'problem' to solve, it is usually viewed as an opportunity for investigation that will lead to a better understanding of an aspect of the discipline. However, when a teacher has a 'problem' it is often viewed not as an opportunity, but as something bad that needs to be dealt with quietly (Bass 1999). The scholarship of teaching and learning movement is working to have 'problems' in teaching viewed as opportunities to improve the state of teaching within individual disciplines, much in the way that research problems are currently viewed. In fact Boyer (1991), one of the initiators of the scholarship of teaching and learning movement, states that 'it is proper to the role of the scientist that he not merely find the truth and communicate it to his fellows, but that he teach, that he try to bring the most honest and most intelligible account of new knowledge to all who will try to learn.' Boyer goes on to say that 'what we urgently need today is a more inclusive view of what it means to be a scholar – a recognition that knowledge is acquired through research, through synthesis, through practice, and through teaching.' Lee Shulman, President of the Carnegie Foundation for the Advancement of Teaching, argues that 'for an activity to be designated as scholarship it should manifest at least three characteristics: It should be public, susceptible to critical review and evaluation, and accessible for exchange and use by other members of one's scholarly community' (Shulman 2000). These are the generally accepted criteria for the scholarship of teaching and learning.

The impetus for an instructor to become a scholarly teacher often begins with an issue in the classroom. Most of us in the sciences start teaching using the methods and practices that we experienced as undergraduate and graduate students, especially

D.M. Eastburn (✉) and C.J. D'Arcy
Department of Crop Sciences, University of Illinois, Urbana, Illinois, USA
e-mail: eastburn@illinois.edu

N.V. Hardwick and M.L. Gullino (eds.), *Knowledge and Technology Transfer for Plant Pathology*, Plant Pathology in the 21st Century 4,
DOI 10.1007/978-1-4020-8934-3_4, © Springer Science+Business Media B.V. 2010

those that we found effective for our preferred styles of learning. However, as new instructors, we often find that the methods that are logical and effective for us are not effective for at least some of the students in our courses. Some instructors who find themselves in this situation blame the problem on the students. Others begin to look at their teaching objectives and practices to determine if there might be things that they can do, as teachers, to better facilitate the process of student learning.

The process of becoming a scholarly teacher involves reflection on one's own teaching and a desire to become better informed about the current state of knowledge of effective teaching practices. This process often follows a pattern of identifying a question or 'problem', searching for a solution, and then conducting a formative evaluation of that solution to determine if it was effective (Bass 1999). The result of this process is usually more effective teaching and increased student learning in the classroom. This often requires a complete re-evaluation of one's objectives for a particular assignment, topic, or course curriculum, something on which many instructors are not willing to invest time and effort.

The scholarship of teaching and learning (SoTL) requires systematic investigation of an instructional problem with the findings made public through presentations or publications. The goal is not only to find solutions to the particular problems in your own classroom but to then make the findings available to other instructors, resulting in improved teaching and learning beyond your classroom, and to subject your findings to critique and evaluation. This knowledge can then serve as a platform on which other teachers can build (Hutchings and Shulman 1999).

4.2 SoTL Projects in Our Classroom

We will present three examples of SoTL studies in an introductory, general education plant pathology course. Plant Pathology 200: Plants, Pathogens and People is an undergraduate general education course that is taught at the University of Illinois at Urbana-Champaign. The 75 undergraduate students who take the course every semester can use it to fulfil requirements in advanced composition and/or natural science. While all of the students have completed an introductory composition requirement, their backgrounds in biology vary greatly as they come from six to eight different colleges or schools and range in class rank from freshmen to seniors. In Plant Pathology 200, students learn about our major crop plants and their most important diseases. These plant diseases are also used to introduce students to issues in agriculture, such as monoculture and genetic diversity, mycotoxins and food safety, pesticides and environmental quality, and genetic engineering and regulation of food production. Although similar courses are taught at other institutions, their number is relatively few, as compared to introductory courses in biology, chemistry, or physics. Because of this, the number of commercially produced teaching resources available for the topics covered in the course is limited. To resolve this problem, over the past 15 years we have created a variety of instructional media and assessment formats to try to ensure that each student has the opportunity to be successful in the course.

As scholarly teachers, we want to know if the resources that we develop and the instructional methodologies that we employ are actually effective in increasing student learning. As faculty members interested in the scholarship of teaching and learning we have carried out studies to allow us to measure and evaluate the impacts of what we do on student learning in our course so that we can improve our own course and pass the knowledge that we gain along for the benefit of other instructors.

4.2.1 Example 1: The Effectiveness of Different Media and Instructional Methods

As in many disciplines, the diversity of students taking our course is increasing. Our classes include second, third and fourth year students from six or more different colleges in the university, and the students have a variety of backgrounds, interests and motivations. In an attempt to accommodate the diversity our students' learning styles, backgrounds, genders, and majors, we present material using multiple instructional formats and several different types of media including standard lectures, videos, podcasts, small group activities, in- and out-of-class writing assignments, a text book, a web-site and student response devices (D'Arcy et al. 2007). We wanted to know if the learning needs of this diverse population of students was being met, and if particular groups of students, based on gender, academic major, or identified learning style, found particular instructional formats or media to be most beneficial.

Over a period of five semesters, students in our course voluntarily completed the Gregorc Style Delineator (Gregorc 1982, 2007) near the beginning of the semester and a survey on preferred media for learning near the end of the semester (D'Arcy et al. 2008). Student perceptions of the effectiveness of the different methods and media were rated using a scale from one (totally ineffective) to five (highly effective). We grouped students by preferred learning style, by gender and by major (science vs non-science) for analyses of the effectiveness ratings. We also conducted focus group interviews with small groups of students after they had completed the course in order to understand better why they found particular methods or media useful or ineffective.

In this study we found that many of the instructional methods and media were perceived as useful by students of all learning styles, majors and genders. It is important to note that these are the students' perceptions rather than independent measures of the usefulness of different methods and media, and student perceptions may differ from what actually impacts their learning. Eight instructional methods (lecture, chalkboard notes, overheads, PowerPoint slides, (clickers), review grids, on-line quizzes and PowerPoint notes) were rated as effective by each of the four learning style groups, by both genders, and by science and non-science majors in two or more semesters. Thus, these media may be generally useful for diverse student audiences. The fact that diverse sets of learners found these methods to be useful is encouraging from a practical perspective, and supports the notion that all students benefit from diverse instructional media and methods.

Our findings challenge views held by some that particular media are uniformly superior to others, or that particular students can learn in only one way, leading us away from a quest for the ideal instructional medium. Instructors should be encouraged to experiment with a variety of instructional media, not only to determine what works, but also to figure out why it is effective. We used these results to reassess our use of particular media, and to make changes so that they became useful to more students or de-emphasised if they were only serving the needs of few. For example, we changed the textbook in our course from 'required' to 'recommended' when it became clear that only a few students found it useful. We believe that the changes we made, as a result of these finding, have significantly improved the quality of learning in the course, and we continue to experiment with new media and methods of teaching and to evaluate their effectiveness for our students.

4.2.2 *Example 2: The Impact of Web Based Assignments*

A second study was undertaken to evaluate effects of a supplemental course web-site (www.ppp.illinois.edu) on student learning. The Plants, Pathogens, and People (PPP) web-site was designed to enhance student learning by providing information in a format that the students could access at times most convenient to them, and to allow them to review the information at their own pace (Bruce et al. 2005). On this site students can obtain in-depth information on some of the important plant diseases covered in the class, as well as conduct virtual laboratory experiments that are designed to reinforce some basic plant pathology concepts through active learning.

The PPP site was developed over a number of years, and we used both summative and formative students' evaluations of the site to expand and improve the site. Analysis of student survey data revealed seven main themes relating to student use of the site: the web allows learner control of information access; the web provides greater learner control of pace of learning; the web holds rich information resources; multimedia possibilities are expanded; students can learn through writing as they investigate the medium; active engagement is encouraged through interactive features such as dynamic simulations and online laboratories; and the web provides opportunities for inquiry and exploring phenomena in depth (Bruce et al. 2005). However, although student feedback indicated that they enjoyed using the site and believed that it improved their learning in the class, we did not have any direct evidence that use of the site resulted in an increase in student learning. To investigate this issue we initiated a study to determine whether the PPP web-based assignments, as a supplement to the lecture material, resulted in better understanding and retention of the topics covered.

In order to determine if student learning was increased by using the PPP site, we evaluated student performance on specially designed questions on the final exam, comparing scores on specific questions among students who completed different disease modules on the site. We found that when students read factual material and completed virtual laboratory activities on the site their comprehension of course material increased somewhat, but improvements were not as consistent or large as we had hoped.

We also conducted focus groups with small groups of students who had recently completed the course to ask them how they used the web-site, and to get their impressions of the usefulness and relevance of the web-based activities. Many of these students indicated that while they found the material and activities on the site to be useful for clarification or review, most of it was just a repetition of information that had been presented during the lecture. Therefore, they only spent as much time on the assignments as was needed to get the grade they wanted. Our assumption that if we provided an interesting, well organised web-site then the students would be internally motivated to use it was not correct.

We felt that the problem was not with the design of the PPP web-site, but rather with ways in which we were asking the students to use it. We therefore decided to restructure the learning objectives for the use of the site, and changed the nature of the web-based assignments to address those objectives more fully. Instead of using the site primarily as a review of material covered during the lecture, we changed the learning objective for these assignments. The new objective is to have students using the site demonstrate an increased ability to describe and apply the scientific method, a topic not covered in the lecture part of the course. The nature of the web assignments were also changed in that the students were asked to conduct virtual experiments, including the generation of hypotheses and conclusions, and the students were then asked to reflect on their experiments using a journal format. We also made the assignments more repetitive in nature with prompt and meaningful instructor feedback before the next assignment was due.

We evaluated the effect of this activity on student learning by comparing scores on pre-test questions relating to the comprehension and application of the scientific method with those of similarly worded post-test questions. The result was a significant improvement in the students' ability to describe the general concepts of the scientific method and to apply that knowledge to specific situations. In this case, altering the learning objectives and the nature of the assignment resulted in greater effectiveness of the site for improving student learning. We found that students put more time and effort into completing the assignments, and that the repetitive, reflective nature of assignments and prompt feedback on student performance reinforced their understanding of concepts being presented.

We found no differences between pre- or post-assignment abilities of students to explain or apply the scientific method associated with student gender or academic major. The fact that female and male students showed equal abilities at the beginning of the course, and showed equal levels of improvement at the end of the course, although gratifying, was not surprising. However, the fact that students in non-science majors and science major showed equal abilities at the beginning of the semester and equal levels of improvement by the end of the semester makes us wonder if our science students are being adequately trained in the scientific method in their other science classes.

4.2.3 Example 3: The Relevance of Course Information to Student's Lives

As teachers, we hope that our students understand the relevance of what they are learning in the classroom and can apply it to their daily lives. But how often does this actually happen? Do students 'take home' knowledge, experiences, and skills and apply them or reflect on them in settings outside the classroom? Do they transfer knowledge and experiences from academic to non-academic settings? If so, how, when, where, and with whom does it happen? And are there things that we can do as instructors to increase the likelihood of it happening? A third study in the Plants, Pathogens, and People course was designed to help us understand better the situations in which students think about and/or apply what they have learned in an academic setting to their non-academic lives.

Over two semesters students in Plants Pathogens and People were asked to record and reflect on instances when they thought about, applied, or discussed with friends or family members information and/or issues presented in the course. Specifically, students were asked to answer such questions as: 'Outside of class, did you think about or talk about anything related to class?', 'If so, what did you think or talk about?', 'What impact did this have on your behaviour in class or outside of class?'. After each semester, student journal entries were coded and evaluated to determine if certain subjects, situations, and/or instructional methodologies were positively or negatively associated with the students using material in non-academic settings.

Analysis of journal entries from students enrolled in PLPA 200 indicated that topics directly related to a student's past and present personal experiences were 'taken home' most often, and this occurred when students were reflecting on course information or were talking with friends or family members. Anything that broke from the usual lecture format had a tendency to interest students and motivate them to discuss the topics outside of class. In many cases these thoughts or discussions arose when the students were eating, preparing, or shopping for food, or when driving by or visiting agricultural settings. The topics most commonly reported tended to relate to class information presented in the form of a story, associated with a 'show and tell' object, associated with a specific visual or related to a writing assignment. Topics directly relating to student experiences, such as coffee, organic produce, and mould growth, were frequently mentioned. In some cases students indicated changes in personal behaviour, such as washing fruit and vegetables, based on information used for a writing assignment.

There follows a few sample excerpts from PLPA 200 student journals:

> Plant pathology is around me more than I thought. I went with my mom to get something to eat, and she got a cup of coffee. I told her the legend of coffee and how it is believed to have been discovered by a boy and goat. I also explained to her how coffee was spread to several places and eventually how coffee rust created a huge problem. I mentioned that this is when coffee production stopped and when tea started to rise in production and popularity.

> This week in Plant Pathology really took me by surprise on the topic of pesticides. There was one slide where all the common misconceptions about pesticides were listed and to my surprise

> they were all what I had always thought about pesticides. That day during dinner I asked my friends about their opinions of what pesticides were and if they thought they were safe.

> This week, while avidly working on my primary macrotheme draft [a writing assignment], I was prompted to discuss the viability of E85 ethanol as an alternative fuel source with my girlfriend. I felt inclined [to discuss why] E85 will never be our alternative fuel answer. At first, she questioned who I was and what I had done with her boyfriend. Being an environmental savvy and green individual myself, I convinced her that E85 was not a true viable option, both environmentally and economically, by showing her various facts presented in the PNAS journal article

We did observe some differences in the subjects of journal entries between semesters. For example, students in the 'Fall' were more likely to write about apples, as many of them visited a local apple orchard when apples were being harvested, whereas students in the Spring were more likely to write about trees once trees started 'leafing out' after winter dormancy.

The student journal entries revealed to us that student use of course information in non-academic settings occurred most frequently for information that had direct relevance to their personal lives, often associated with food, or pertaining to issues currently in the news. The impact of information delivery through story telling was especially notable, as was the impact of out-of-class writing assignments ('writing to learn'). One benefit of addressing a topic by telling a story is that the information is organised in a format that is easy for the student to recall later and retell to others. One of the goals of the course is for students develop a better understanding of how agriculture in general, and plant diseases in particular, have impact on their everyday lives. The results of this study validate our use of story telling, topical writing assignments and current events in the news, and our highlighting of the personal relevance of topics for the students as ways to achieve this goal. We also plan to look for additional aspects of the course that can be presented in these manners in order to increase the likelihood that the students will use the course information in non-academic settings.

4.3 Why Do SoTL?

One question that we are frequently asked by colleagues and administrators alike is 'Why should we do SoTL research?' After all, devoting time to a SoTL project does take time away from other important activities, such as disciplinary research. There is currently not a lot of prestige associated with developing an expertise in SoTL research, and, though it is improving, there are not a lot of resources available to fund SoTL projects. Another criticism is that we do not have the expertise needed to conduct meaningful research projects in the area of teaching and learning, and that we should leave such research to those in the fields of education and psychology.

One reason that faculty members should be interested in SoTL research is to improve their effectiveness as a teacher. A second reason follows from the first, to increase the level of student learning that goes on in (and out of) the classroom.

Taking the time to challenge assumptions and to investigate what is happening at the student level can result in significant improvements in student learning. In a recent article in *Science* a physics professor at Harvard describes the process by which he determines that his lectures are not very effective, and how he completely restructures his teaching methodology, using current information on learning, to greatly enhance student learning in his classroom (Mazur 2009). By abandoning the traditional lecture format he is able to engage his students more completely in the learning process. A third reason for plant pathologist to do SoTL research is to improve the state of teaching in the discipline of plant pathology, if not the greater academic community. If plant pathologists do not endeavour to improve teaching practices in plant pathology courses, who will?

A few years ago, during an open discussion at an international meeting on SoTL, a prominent and well respected expert on educational research stood up and told those assembled that they should stop playing around and leave research on education to the experts. Not being one who is quick with a comeback, I figured out how I should have responded to that comment several hours after the session had ended. What I should have reminded the group was that the development of the force concepts inventory, to determine if students in introductory physics courses are truly developing an understanding of the Newtonian concepts of motion and force, resulted from physics instructors determining that while their students scored well on assignments and exams, when questioned more rigorously it became clear that they were not replacing previously held misconceptions. It was these physics instructors, not educational theorists, who determined that there was a problem, investigated the cause, and redesigned their teaching objectives and methods to solve the problem. Educational theorists do not know what specific problems we are facing in our classrooms because they are not trained in our disciplines and do not have the expertise to determine whether or not students are learning the correct information and concepts. It would be difficult for someone who is not trained in plant pathology to determine if students in our classes are learning how to interpret properly a diagram of a disease cycle, or to understand the differences between a monoculture cropping system and genetic uniformity of a host crop.

While education researchers and theorists have developed a great deal of information that is relevant to our classrooms and can be used to improve our effectiveness as teachers, these people do not have the time or inclination to adapt their information for teaching to all of the disciplines in which it may be useful. This situation is somewhat analogous to the differences between those who do basic research *vs.* those who do applied research in the discipline of plant pathology. Those doing basic research develop new information and new theories using model systems, but it is the applied researcher who takes that information and adapts, for example, basic research on *Arabidopsis* to develop practical disease management practices on apples or soybeans. And it is the applied researcher who identifies problems that may lead to interesting new lines of inquiry for basic researchers. Similarly, while those of us teaching within our disciplines can benefit from the work of educational theorists, we can also identify the problems that lead to new areas of educational inquiry. In our SoTL work, we often collaborate with colleagues in departments of education or information science. We benefit from their expertise in the development of effective research methodologies and their insights into the broader meanings of our research findings. They, in turn, appreciate the opportunity to apply their expertise and test their theories in real world settings.

A final benefit to having plant pathologists engage in SoTL research is the increased visibility of plant pathology. As a result of conducting research in our classes and then presenting our findings at teaching seminars and workshops on campus, or as presentations and posters at teaching related professional meetings, many more people have now heard about the discipline of plant pathology and are at least partially aware that plant diseases affect their lives daily. Although the challenges are large and the rewards are often small, we believe that the benefits of doing SoTL research within the discipline of plant pathology definitely more than justify the effort required.

References

Bass R (1999) The scholarship of teaching: what's the problem? Inventio 1:1. http://www.doiiit.gmu.edu/Archives/feb98/randybass.htm

Boyer EL (1991) The scholarship of teaching: from "scholarship reconsidered: priorities of the professoriate". Coll Teach 39:11–13

Bruce BC, Dowd H, Eastburn DM, D'Arcy CJ (2005) Plants, pathogens, and people: extending the classroom to the web. Teach Coll Rec 107:1730–1753

D'Arcy CJ, Eastburn DM, Bruce BC (2008) How media ecologies can address diverse student needs. Coll Teach 55:1–7

D'Arcy CJ, Eastburn DM, Mullally K (2007) Effective use of a personal response system in a general education plant pathology class. Plant Health Instr. doi:10.1094/PHI-T-2007-0315-07

Gregorc AF (1982) An adult's guide to style. Gabriel Systems, Maynard, MA

Gregorc AF (2007) Gregorc style delineator. http://gregorc.com/instrume.html. Accessed 8 Jan 2009

Hutchings P, Shulman LS (1999) The scholarship of teaching: new elaborations, new developments. Change 31:10–15

Mazur E (2009) Education. Farewell, lecture? Science 323:50–51

Shulman LS (2000) Fostering a scholarship of teaching and learning. Institute of Higher Education, Georgia University, Athens

Chapter 5
Technology Transfer in Extension: Experience in the United States of America

P. Vincelli

5.1 The Impact of the Internet

5.1.1 Real-Time Updates

Being up-to-date is possible in a way that never was before. For example, the pathogen that causes Asian soybean rust (*Phakopsora pachyrhizi*) was introduced into the United States of America (USA) in 2004, most likely in winds associated with Hurricane Ivan. Through the use of internet-based resources, maintained by a consortium of scientists from multiple institutions, soybean producers have been able to stay up-to-date throughout each growing season from information on where the pathogen is and whether or not fungicide application is warranted (http://sbr.ipmpipe.org/cgi-bin/sbr/public.cgi). This web resource and affiliated extension programmes have helped avoid millions of dollars worth of unnecessary fungicide applications, as well as helping growers prevent losses in those instances when the disease was threatening crop yields. The high level of co-ordination, communication, and education that have benefited soybean producers would have not been possible without the internet.

5.1.2 Readily Accessible Training Materials Available '24/7'

Educational materials of all sorts are available instantly to growers: publications, digital images, downloadable PowerPoint presentations (which can include recorded narration) and instructional video clips. Remote access to live PowerPoint presentations from personal computers is becoming increasingly commonplace in the USA.

P. Vincelli (✉)
Department of Plant Pathology, University of Kentucky, 207, Plant Science Building, Lexington, KY 40546-0312
e-mail: pvincell@uky.edu

N.V. Hardwick and M.L. Gullino (eds.), *Knowledge and Technology Transfer for Plant Pathology*, Plant Pathology in the 21st Century 4,
DOI 10.1007/978-1-4020-8934-3_5, © Springer Science+Business Media B.V. 2010

Although these 'webcasts' (also called 'webinars') are valuable, they have not supplanted extension conferences but rather simply added a new vehicle for information dissemination. These developments not only provide convenience to producers who have internet connections; they also help to promote efficiency and maintain extension productivity in times of generally static to declining budgets.

5.1.3 Quality-Assured Web Resources

Google searches yield an overwhelming array of 'hits', leaving users to distinguish sound information sources from biased or low-quality sources. Since some users cannot make these distinctions, in recent years there have been efforts to launch web sites providing agricultural science-related information that is all quality-controlled.

*e*Xtension (www.extension.org) has resulted from a partnership of over 70 universities as a vehicle for extending diverse programmes of land-grant universities, including those relating to agricultural science. All publicly accessible information is peer-reviewed to assure that it is objective and research-based. Plant pathology is still a minor component of this nascent effort, since much of the initial work has related to animal production, but the discipline will be expected to be a significant component of crop-related information as that dimension continues to grow. Information available through *e*Xtension is free.

The purpose of the Plant Management Network (PMN) (www.plantmanagement-network.org/) is also to provide quality-assured scientific information, and plant pathology is a major component of this effort, given PMN's association with The American Phytopathological Society (APS). Scientists from land-grant universities and other organisations provide and review content, including research papers and reviews, on-demand webcasts, digital images, and extension publications. Access to PMN is by subscription.

5.1.4 Digital Diagnostics

Internet-based tools have been developed both by land-grant universities and certain commercial interests that use digital imagery as the basis for plant disease diagnosis, without necessarily examining an actual plant specimen. This practice was initially very controversial, since there was disagreement about how accurate a diagnosis could be without examination of diseased plant material by a professional plant pathologist. There has been some evolution of this practice since its introduction during the previous decade. Some diagnostic programs receive only digital images of host symptoms and field distribution, and use that information only for discussion/consultation purposes in the absence of an actual plant sample. In those laboratories, if an actual sample is provided, the digital images are used as supplementary information to help in making a diagnosis. Other diagnostic programs use custom-designed web-based digital image systems to receive images not only of host symptoms but also images resulting from microscopy, and they can sometimes confidently issue diagnoses based on these images. The initial controversy has subsided probably because diagnosticians essentially agree about the limitations of 'digital diagnostics'. There seems to be general agreement that conclusive evidence of the presence of the pathogen is essential for proper diagnosis, and if that evidence is available in the digital images provided, a positive diagnosis may be possible. In contrast, if that evidence is lacking from digital images provided, or if further tests (ELISA, PCR, additional microscopy) are needed, then a proper diagnosis cannot be issued based on digital imagery alone.

5.2 The Changing Role of the Extension Specialist

At one time, university-based extension specialists were the gatekeepers of the latest information from the agricultural sciences. Today, information flow is no longer vertical, from researchers through the state and local extension system to farmers. Producers now receive information from a wide range of sources: seed companies, fungicide manufacturers, crop consultants, fertiliser suppliers, extension materials from other states, published research reports. For most producers, extension is only one of many sources for their day-to-day information needs, and often is not even be the most important one.

If extension is to continue to be relevant in this new world, extension specialists must protect their credibility as unbiased, science-based experts, by assuring that they are always living up to that standard in everything undertaken. Although producers obtain information from many sources, extension specialists hold a unique place in providing applied research and recommendations widely regarded as unbiased, as well as providing the most advanced diagnostics.

Nucleic acid-based techniques are becoming fundamental to extension programmes in the USA, increasing precision and speed of pathogen detection and identification. More and more applied plant pathology programmes are incorporating molecular tools for pathogen detection and diagnosis, and in some instances (for example, sudden oak death caused by *Phytophthora ramorum*), detection by nucleic acid-based tools is required for completion of the diagnosis. As valuable as these tools are, it is important to remember that appropriate use of molecular tools requires awareness of their limitations as well as strengths. These issues were addressed in a recent review paper on nucleic acid-based pathogen detection (Vincelli and Tisserat 2008).

5.3 Evolving Extension Programming

In recent decades, extension programming in the USA has diversified well outside the boundaries of production of food and fibre. Educational programming has emerged to address widely diverse issues, including water quality, food safety, invasive species, farmers' markets, improving community health, disaster education, energy efficiency, alternative energy sources and even art. Although plant pathology plays a role in some of these areas, it has no technical role to play in many others. Expansion of extension to address societal issues well outside of agriculture is a positive development but it does mean that plant pathologists need to continue to perform well, so that their relevance to society continues to be recognised. The emergence of issues-based programming, while a positive development, may also contribute to the decline in importance of extension to local producers, since it is very difficult for agents to have educational programming on widely divergent issues while still maintaining expertise in agricultural production.

As state and federal funds for extension programming have declined, funding from competitive grants and local sources has become more important. Funding cuts have also resulted in creative solutions to filling existing needs, such as instances where an extension specialist is hired with a joint appointment split between two neighbouring states. Some states have consolidated county extension programmes, so that there is no longer an extension worker in each county who addresses educational issues relating to agriculture. Though difficult budget choices must sometimes be made, this has led to a decline in visibility of extension in agricultural communities.

5.4 Privatisation

It is important to be reminded that it is legitimate to ask whether, in a capitalistic society, the services plant pathologists perform in extension would not be more appropriately done by a for-profit, private enterprise. However, extension specialists provide important services for society: namely, providing objective applied research, unbiased disease control recommendations, and advanced diagnostics for our crops, landscapes, and forests. Fortunately, this seems to be appreciated by society, at least in halls of state and federal legislators. There have been no serious discussions regarding privatisation of extension services in the USA. Such a move would isolate extension from the land-grant university and would work to the detriment of both

Reference

Vincelli P, Tisserat N (2008) Nucleic acid-based pathogen detection in applied plant pathology. Plant Dis 92:660–669

Chapter 6
Diagnostic Networks for Plant Biosecurity

J.P. Stack

6.1 Why Plant Health Is Important

Plant systems provide an array of ecosystem services that support the survival of most species, including humans. Plants are important generators of oxygen and consumers of carbon dioxide. They contribute to the purification of both water and air. Plants provide shelter directly for wildlife species and indirectly for humans through the harvest of trees for timber used in the construction of houses. The fibres of plants are used in many ways, for example in clothing and ropes. The well-being of humans is dependent upon the health of plant systems.

6.1.1 Food

Plant systems are the foundation upon which public health and food production systems are built. The complex nature of food production and distribution systems, their dependency on the supply and economics of energy, and their susceptibility to political instability, make sustainable food security a significant challenge (Anon 2008). Public health and food security will be difficult to achieve and more difficult to sustain if the population at risk is undernourished and at risk of starvation (Audibert et al. 2003). Because of the relationships among food prices, fuel prices, standard of living, environmental health and international trade, global food security for our expanding population will depend upon increasing food production significantly (Collier 2008). A United Nations Millennium Goal is to reduce poverty by 2015 (Anon 2008). 'Hunger is both a cause and consequence of poverty' (FAO 2001). The vast majority of the world's population is directly dependent upon

J.P. Stack (✉)
Department of Plant Pathology, 4024 Throckmorton Plant Sciences Center, Manhattan, Kansas, 66506, USA
e-mail: jstack@ksu.edu

N.V. Hardwick and M.L. Gullino (eds.), *Knowledge and Technology Transfer for Plant Pathology*, Plant Pathology in the 21st Century 4,
DOI 10.1007/978-1-4020-8934-3_6, © Springer Science+Business Media B.V. 2010

plant-based foods as the primary source of calories to sustain life. There is a strong correlation between the standard of living within a nation and the consumption of protein derived from animals. As a nation's standard of living increases, there is a concomitant shift from a primarily plant-based diet to a meat-based diet. The challenge this presents is that it takes several kilograms of plant-based feed to generate 1 kg of meat for direct consumption, for example it takes 6 kg of grain to generate 1 kg of beef (Collier 2008). Consequently, nations with a higher standard of living and higher intake of meat-based protein are even more dependent upon healthy and sustainable plant systems. As we push to raise the standards of living of the world's population at the same time that the population is expected to increase by almost 50% over the next 25 years, our plant systems may be pushed beyond their current capacity to meet the demands.

Diamond (1997) made a strong case for the causal relationship between the rise and success of many civilisations and the development of productive plant and animal agricultural systems. He made an equally compelling argument that many civilisations and societies declined to insignificance as a consequence of failing to protect their natural resources, including their plant systems (Diamond 2005). As the world's population increases and poverty is diminished, there will be unprecedented pressure on the world's plant-based food production systems to satisfy the demand; every kilogram of food and feed will matter. We must invest in the science and technologies that will increase food production; we must also invest in the science, technologies and policies that will protect the food that is produced.

6.1.2 Medicine

Many of the medicines used to treat diseases and heal people from physical and mental afflictions are derived from plants. As population growth, environmental degradation, landscape exploitation and climate change put enormous pressure on plant systems over the next century, the availability of plants that provide the natural medicines for indigenous people as well as the preservation of plant species for the discovery of new drugs will decline significantly.

6.1.3 CO_2 and Sequestration

Plants will play a vital role in managing increases in global atmospheric carbon dioxide concentrations for the foreseeable future. Elaborate carbon credit and exchange systems are being developed that utilise plants' ability to sequester carbon as a means to offset the carbon emissions of the industries that support the global economy. Both foliar and root diseases can reduce the effective photosynthetic area of plants, thereby limiting the amount of carbon assimilated by the plant. As epidemics progress and the amount of photosynthetic tissue decreases, the affected plant population will shift from a carbon sink to a carbon source releasing carbon dioxide as a product of tissue decomposition. To maximise the carbon capture benefit, plant systems must be kept healthy. Within any plant growth cycle, epidemics must be prevented or their progress slowed to delay the transition from carbon sink to source.

6.1.4 Fuels

Plant-based fuels are fast becoming an alternative to fossil fuels. Although the long-term economic viability of plant-based fuels is yet to be determined, many nations are exploring these technologies to reduce their carbon footprint and to minimise the economic and environmental burden of dependency on fossil fuels. Healthy plant systems will be necessary if plant-based fuels are to become a viable option for mitigating the negative effects of climate change without compromising food security.

6.1.5 Aesthetic Value

Across the globe and throughout history, humans have valued plants for their aesthetic value. We care about the appearance of natural and managed landscapes. Plants are part of our culture. Global trade in ornamental and landscape plants is a multi-billion dollar (US$) industry contributing to the local and national economies of many nations.

6.2 Challenges to Sustainable Plant Health

There are many challenges to plant health including the threats posed by global trade, climate change, the expanding global population, bioterrorism and biocrime (Mack and Lonsdale 2001; Stack 2008; Stack et al. in press). These challenges will require investments in research and technology development to protect the plant systems that support life and stabilise ecosystems. Two of the most pressing problems are global trade in plants and plant products and population growth and development. The relative significance of intentional introductions of plant pathogens is difficult to assess and has been the subject of much discussion (Cochrane and Haslett 2002; Cupp et al. 2004; Madden and Wheelis 2003; Waage and Mumford 2007; Wheelis et al. 2008; Whitby 2002; Young et al. 2008).

6.2.1 Global Trade

Tremendous volumes of plants and plant products are transported across oceans, continents, and national boundaries every day. Pests and pathogens are commonly comingled with the plants and plant products greatly increasing the risks of introducing organisms with the potential to damage natural and managed ecosystems (Britton 2004; Campbell 2001; Palm 1999; Rossman 2001). Modern transportation systems make possible the shipment of large quantities of plant material over great distances in short periods. In many cases, the time taken to ship plants from one hemisphere or continent to another is less than the incubation period for many diseases. This reduces the effectiveness of visual inspection as a means to intercept pathogens and pests on plants or plant products and consequently increases the risk of unwanted introductions. The magnitude of international plant trade precludes prevention as a sole strategy for protecting plant systems; it cannot all be inspected. In addition to a prevention strategy, each nation needs the infrastructure to detect rapidly the introductions that elude the inspection, interception and quarantine programmes.

In addition to the movement of previously described pathogens, another unintended consequence of the large-scale movement of live plants is the emergence of new species of plant pathogens as the result of the hybridisation of two species that historically would not have come into contact with each other (Brasier 2008; Man in't Veld et al. 1998). In one instance, the hybridisation event was made possible only as a direct result of global trade in landscape plants (Brasier 2001). Of particular concern is that the host range and virulence patterns of the new hybrid species could not have been predicted from the parental phenotypes.

6.2.2 Population

At present, approximately one sixth of the world's population lives in poverty and hunger. Food protests and riots in many nations over the past year demonstrate the fragile nature of global food security (Shelburne 2008). Population models predict growth of almost 50% by 2050. This growth will mostly occur in urban, non-food producing environments. A major problem is ignorance of food production in the developed nations where the world's wealth is concentrated and from where the resources to address these problems must come. Few are aware of the infrastructural requirements necessary to ensure the adequate supply and timely distribution of food. As the world's populations become increasingly more urban in nature, their connections to food production are weakened. The literal and figurative distance between people and agriculture may adversely influence the policies affecting agricultural systems. To satisfy the food demands of the future we will need to increase production of plant-based foods and feeds and to protect that production from losses during transport and storage. Population growth will put enormous pressure on plant systems as a result of competition between land for production and land for development.

6.3 The National Plant Diagnostic Network

6.3.1 Mission and History

In the United States of America (USA), agriculture was identified as an integral component of the nation's critical infrastructure. Vulnerability assessments identified weaknesses in our ability to protect agricultural systems from the increasing threats from natural introductions of pests and pathogens associated with trade and from the emerging threats of bioterrorism and biocrime. Among the vulnerabilities identified was the decreased capability and capacity in plant diagnostic laboratories nationwide. A decades-long decline in funding to support state plant diagnostic laboratories resulted in a wide discrepancy from state to state.

In 2002, the United States Department of Agriculture's Cooperative States Research, Education and Extension Service established the National Plant Diagnostic Network (NPDN) in collaboration with the Land Grant University System to help protect plant systems from the intentional, accidental, and natural introductions of plant pathogens and insect pests (Stack et al. 2006). The outbreak of a disease can be represented as a simple process model based on the critical system components and the sequential stages from the introduction of the pathogen or pest through the eventual resolution of the outbreak (Fig. 6.1). A plant biosecurity system must have strategies prepared for each stage of the outbreak cycle; a

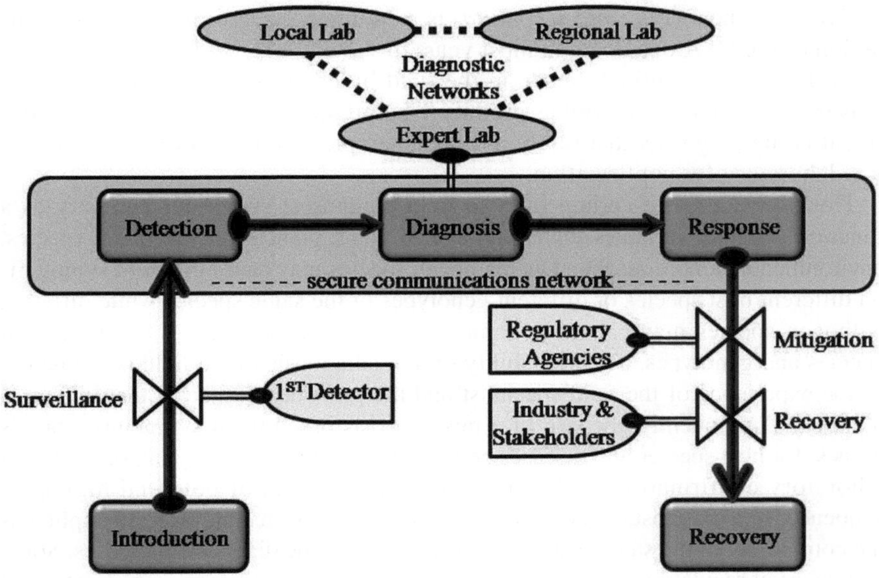

Fig. 6.1 A disease outbreak model commencing at the introduction of a pathogen into a plant system to the resolution of the disease outbreak. Regulators of key processes are identified

strategy to minimise the risk of introductions, a strategy to maximise the probability of early detection of an outbreak, a strategy to ensure the rapid and accurate diagnosis of the new invasive agent, a strategy to ensure the effective management of the outbreak and a strategy to facilitate a successful resolution of the outbreak. NPDN was established specifically to address the early detection and the accurate diagnosis of the causal agents associated with disease outbreaks. The initial emphasis was placed on developing the infrastructure and training necessary to achieve a state of preparedness for a small list of high consequence plant pathogens termed 'Select Agents'. 'Select Agents' were identified as those pathogens presenting a very high risk to plant systems in the USA and that could be weaponised for intentional introductions. However, from the inception, NPDN was designed to be of value in protecting plant systems from all types of introductions; accidental, natural or intentional.

6.3.2 Diagnosis Hierarchy

In the overwhelming majority of cases, the diagnoses of routine diseases are often based upon the signs and symptoms present at the time of diagnosis. This is true for the diagnosis of disease in plants, animals and humans. A priori knowledge of the system, host and history of the area are important aids in the success of these diagnoses. In many cases, diagnoses are made in the field when the disease is common to the area and familiar to the diagnostician. For example, leaf rust of wheat (*Puccinia recondita*) is a common disease in the Great Plains region of the USA occurring in most years to varying degrees. Accurate diagnosis of leaf rust is routinely made in the field by extension agents, industry and university specialists, as well as producers based upon symptoms alone. So confident is the diagnosis that rarely are plant samples with this disease submitted to a laboratory for confirmation.

However, there are certain risks to field diagnoses: symptoms can vary as a function of many variables including, age of plant, plant genotype, plant species, environmental conditions, etc. One pathogen species may cause different symptoms on different host species or different genotypes of the same species while different pathogen species may cause the same symptoms on the same or different host species and genotypes. The probability of a misdiagnosis in the field is a function of the experience of the field specialist and the prior occurrence of the disease in the area. Consequently, the risk of a misdiagnosis based upon symptoms alone is highest for host species and diseases new to an area. It is extremely important to get laboratory confirmation of diagnoses for any diseases of potential high consequence. In a diagnostic laboratory, increasing levels of technology are applied as the complexity of the sample and the consequence of the diagnosis increases. Some pathogens of regulatory concern are very closely related to indigenous, non-regulated pathogens. In some cases it can be a single sub-specific taxon, for example *Ralstonia solanacearum* race 3 biovar 2. The danger of false negatives and false positives can be significant with serious consequences of a misdiagnosis. Very detailed molecular diagnostic protocols are often the only means for the accurate identification of regulated pathogens. These molecular protocols are often very expensive and difficult to successfully execute in the field.

6.3.3 Diagnostic Infrastructure and Operations

The disparity among laboratories across the nation, with respect to core competencies and diagnostic infrastructure, required a detailed assessment of capability and capacity of every plant diagnostic laboratory. This assessment provided the data necessary to develop a national plan of work to enhance the national infrastructure. NPDN laboratories were provided with the funding necessary to establish minimum capabilities for routine diagnoses as well as the capabilities to serve as a triage laboratory in a national system.

6.3.3.1 Triage

During an outbreak, the number of samples to be processed can increase rapidly to levels that often overwhelm the human and physical resources of a single laboratory. High consequence samples with regulatory implications must be processed by a laboratory with the authority to make a confirmatory diagnosis with legal standing. NPDN now serves as a support system for the official laboratories that provide the confirmatory diagnoses that underlie regulatory actions and response plans. The concept of triage is one of separating positives from negatives. In concept, the triage process begins in the field at the outbreak site when the initial assessment of the nature of the problem is made (Fig. 6.2). Increasing levels of scrutiny, and often technology, are applied as the samples pass through a diagnostic laboratory and then to an expert laboratory for confirmatory diagnosis. The primary function of the triage laboratory is to identify the negatives and send the presumptive positives to the confirmatory laboratory (Fig. 6.2). This triage process allows the official laboratory to focus on only those samples with the highest probability of being of regulatory concern. This speeds up the diagnostic process facilitating the most rapid response. The decision to stop or to allow the commercial shipment of plants or plant products is dependent upon the confirmatory diagnosis. The economic impacts associated with delayed decisions can be significant.

6.3.3.2 Telemedicine System

The majority of plant diseases are caused by fungi that, in some circumstances, produce structures that can be diagnostic. Many NPDN laboratories are equipped with web-enabled microscopy and video conferencing. This provides the infrastructure

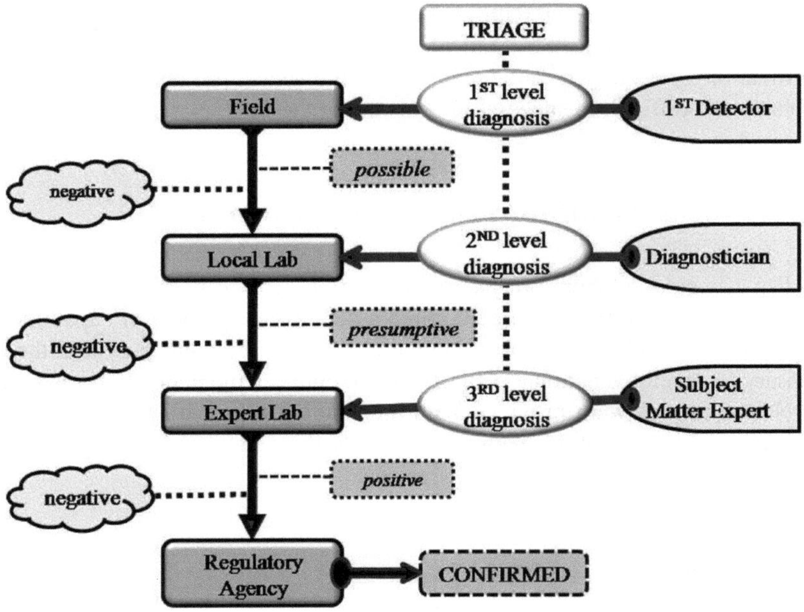

Fig. 6.2 Sequential diagnoses with increasing host or pathogen-specific expertise and increasing application of technology form the basis of a diagnostic triage system

for a functional telemedicine system whereby a diagnostician in the laboratories can send an image of the specimen being viewed with the microscope to a miniserver. An expert for that pathogen or host is then contacted (e.g. by email or telephone) and provided with the URL to a site on the microscope server. The expert at any remote location can view on their computer monitor the image from the microscope in 'real time'. Through video and audio conferencing, the expert and the local diagnostician can reach a collaborative diagnosis.

This system is also used for collaborative diagnostics based on plant symptoms. The imaging systems and web connectivity allows for the real time observations of plants or plant parts by diagnosticians and experts to make a quick decision regarding appropriate diagnostic tests and protocols. This has been very helpful in providing new diagnosticians with little experience access to experienced diagnosticians and specialists.

6.3.3.3 Molecular Diagnostics

An important goal of NPDN is to provide a rapid and accurate determination of cause. Traditional diagnostics in plant clinics with limited access to modern technologies rely on morphological and cultural characteristics of the pathogen. This presents a few potential problems: (1) morphological and cultural characteristics are not always reliable diagnostic determinants due to similarity among different pathogen species, (2) some pathogens of concern are sub-specific taxa that cannot be distinguished morphologically or culturally from indigenous pathogen species, and (3) for many pathogens it can take a long time to derive a diagnosis based on morphological and cultural characteristics. This can delay the implementation of appropriate mitigation measures. For bacteria, it may take a few to several days to isolate and purify a culture for identification. For disease caused by fungi, it may take one to a few weeks to isolate and purify a culture for identification. Meanwhile, at the outbreak site the pathogen is continuing to colonise plants and reproduce resulting in the spread of the disease from the outbreak site. In addition, many plant pathogens are not easily cultured or identified using traditional laboratory methods based upon pathogen morphological and cultural characteristics.

To shorten the time it takes to derive plant disease diagnoses and to increase the accuracy of those diagnoses, NPDN in collaboration with the USDA Animal and Plant Health Inspection Service (APHIS) began a programme for the development, validation and deployment of molecular capability and competence in Network laboratories. Some NPDN laboratories already had the required technologies and expertise while others did not. With a combination of NPDN funding, grant funding and support from host institutions, NPDN laboratories were provided the equipment (e.g. thermocyclers, real time thermocyclers, enzyme-linked immunosorbent assay [ELISA] plate readers, gel documentation systems, etc.) to permit modern nucleic acid-based diagnostics (e.g. polymerase chain reaction [PCR] and real time PCR).

6.3.3.4 Lab Accreditation System

For this laboratory network and triage system to be most effective, the results generated by each laboratory must be of a certain quality to engender confidence in the results and to permit the interpretation of shared information. To meet this expectation, a national laboratory accreditation system (LAS) is under development. The LAS will establish minimum standards of laboratory capability and diagnostician competence. Capability will be assessed based upon the technologies utilised in that laboratory. For a number of reasons, including the wide disparity in plant systems supported, not all laboratories have the same technologies. Competence will be assessed based upon the technologies and protocols utilised by the laboratory and the adherence to the LAS quality manual of operations. The LAS will monitor compliance for all NPDN laboratories.

The LAS will be designed to synchronise with an APHIS managed system for protocol certification of laboratories and diagnosticians. To enhance further the triage system, APHIS is developing a protocol certification system whereby diagnosticians in NPDN laboratories can establish competence in the execution of APHIS-approved standard operating procedures (SOPs) for high consequence pathogens. Diagnosticians are required to complete successfully a proficiency testing programme for each SOP/pathogen. Some of these laboratories will be authorised to provide confirmatory diagnoses with legal standing. This national network of NPDN accredited laboratories in concert with the APHIS protocol certification system enhances both capability and surge capacity.

6.3.3.5 National Repository of Diagnostic Data

A national repository of diagnostic data from all NPDN laboratories was established at Purdue University. This data repository serves as a source of the prevalence and geographic distribution of common pathogens. It also provides a database for epidemiological analyses on disease trends and outbreak dynamics. In the event of an intentional introduction, the data repository will provide the background data for forensic analyses (Fletcher et al. 2006; Fletcher 2008a).

6.3.3.6 Ring Testing and Research Support

The implementation of a network of plant diagnostic laboratories has had impact beyond the original intent. NPDN laboratories have been utilised by USDA research laboratories in a ring testing format to accelerate the validation process for new diagnostic protocols (Lamour et al. 2006). This has added value to the protocol validation by giving the diagnosticians advanced training with the protocol prior to its adoption as a standard operating procedure. In addition, NPDN diagnostic laboratories have been called upon to support research projects associated with disease outbreak response efforts, for example comparative analyses of different diagnostic technologies under actual outbreak conditions (Bullock et al. 2006).

6.4 Communications Infrastructure and Operations

Some of the diagnostic information generated in NPDN laboratories has the potential to affect commerce in either a positive or negative manner. The detection of a regulated pathogen or pest could result in interstate or international restrictions on the importation or export of plants and plant products. It is imperative that such information be secured from inadvertent disclosure prior to a confirmatory diagnosis and an official response decision by state and federal regulatory agencies. NPDN established a secure communications system that allows for the exchange of information among diagnosticians and between diagnosticians and regulatory agencies (Stack and Baldwin 2008). In some states, the communication system is deployed to the county level allowing extension agents and staff to enter preliminary information including digital images of the affected fields, plants, insect pests, and plant parts. The communication system is designed and managed to maintain data integrity and accessibility against a range of natural (e.g. extreme weather events) and human (e.g. vandalism) threats as well as to prevent unauthorised physical or electronic access to Network information (Stack and Baldwin 2008). Communication protocols are tested monthly to ensure that communications are received in a timely manner by those in positions of responsibility and authority.

NPDN maintains a web portal to provide general information for the public and serve as a gateway to the web sites of each regional network. The regional websites are utilised in different ways and to different degrees by each region. For example, the Southern Region website is a very dynamic site used for active communications while the Great Plains Region website is used primarily as a repository of information (e.g. diagnostic SOPs, training and education materials, plans of work) shared by the member states.

6.5 Training and Education

6.5.1 First Detectors

To ensure the early detection of disease and pest outbreaks, first detectors must be aware of the possible threats in their area and the proper procedures to follow should they encounter a field or plant sample that is suspect for a new pest or pathogen. A first detector may be a farmer, extension agent, consultant, agricultural industry professional, homeowner, or landscape manager. The first detector is that person most likely to encounter an outbreak of a pest or a disease in any natural or managed plant system. In the USA, a first detector is distinct from a first responder who may have legal authority for taking action when a new organism is detected. NPDN developed and deployed a first detector training and education programme. It is based on a national curriculum that centres on a core set of training modules that can be delivered over an array of training methods. NPDN training workshops are delivered in face-to-face sessions in the field or in the classroom. An on-line first detector training system is also operational allowing for asynchronous programme delivery. The core NPDN curriculum covers the mission of NPDN, the art and science of diagnosis, sample acquisition and shipping, the use of digital sample submission, and an increasing list of modules dedicated to specific high consequence pathogens and insects.

First detectors are trained to identify the signs and symptoms associated with diseases and damage from high consequence pathogens and insects. They are made familiar with the proper protocols for notification of authorities when a suspected high consequence disease is detected or an otherwise unknown is encountered. An accurate diagnosis is in part dependent upon the quality of the sample received by the diagnostic laboratory. To ensure that the highest quality samples are submitted to the laboratory, first detectors complete a training module detailing the proper procedures for sample collection and shipping to maintain the integrity of the sample and minimise the risk of dispersal of the agent. The importance of secure communications during the time from first detection through regulatory response is also explained.

6.5.2 Diagnosticians

To ensure the rapid and accurate diagnosis of diseases and pests, plant diagnosticians must be trained in the use of advanced diagnostic technologies, the latest SOPs, and the importance of secure communications. NPDN diagnosticians must be familiar with the proper protocols for notification of authorities, the laws and regulations regarding the handling, storage, transport of select agents, sample processing and destruction. NPDN offers regional hands-on workshops on a variety of topics including, diagnostic technologies (e.g. PCR), pathogens (e.g. wheat streak mosaic virus), insects (e.g. *Maconellicoccus hirsutus*), and vectors (*Aceria tosichella*). For many of these, a manual of relevant information is provided to the participants. The telemedicine system previously described has been used in internet-delivered training programmes with great success. APHIS's Center for Plant Health Science and Technology (CPHST) provides hands-on training workshops for all NPDN diagnosticians. These workshops give NPDN diagnosticians the opportunity to gain experience with the most up to date protocols for the pathogens of greatest concern.

As new pathogens and pests emerge and as new technologies improve our efficiency and accuracy in diagnosis, training will be required to keep NPDN diagnosticians prepared and at the forefront of plant biosecurity. Through the collaborative efforts of NPDN, CSREES, and APHIS, a meaningful professional development programme for plant diagnosticians will emerge. This will be an important component of the national plant biosecurity infrastructure.

6.5.3 Preparedness Exercises

The successful management of a plant disease outbreak requires the co-operation of many individuals and organisations and the co-ordination of all the activities associated with reporting, response, and recovery. This involves many individuals representing local, state, and federal agencies, land grant universities, industries, and the owners and managers of the area affected. It is important that, as an outbreak unfolds, everyone understands their role and responsibilities, as well as the roles and responsibilities of everyone involved. It should not be assumed that everyone understands this process.

NPDN developed and implemented a preparedness exercise programme to help states prepare for a plant disease outbreak. Each exercise is based on a scripted plant disease or insect pest outbreak scenario that describes in great detail the sequence of actions and communications that are required for the successful response to and resolution of an outbreak. Scenarios have included simulated outbreaks of Asian soybean rust (*Phakopsora pachyrhizi*), emerald ash borer (*Agrilus planipennis* or *Agrilus marcopoli*), and others. Each exercise is managed with a computer software program that provides all the information necessary for in-depth post exercise analyses.

The end result is a very efficient process for testing the systems necessary for the successful resolution of outbreaks.

6.6 A Case for International Networks

The elevated risks associated with the large-scale movement of plants and plant products around the world make international co-operation a necessary prerequisite to global plant biosecurity (Stack and Fletcher 2007). Plant diseases are a major constraint to food security globally (Strange and Scott 2005). The pests and pathogens indigenous and recurrent in one country are the exotic pests and pathogens of concern to another country. This is certainly true for trading partners where pathways for introduction of pathogens and pests are linked to the vehicles of trade and the volumes of trade preclude complete and detailed inspections of imported plants and plant products.

International networks should be established that link the plant diagnostic laboratories of trading partner nations, nations with common boundaries, and nations within natural pathways (e.g. the *Puccinia* [http://www.ars.usda.gov/Main/docs.htm?docid=11301] and *Peronospora* [Main et al. 2001] pathways of North America). Such networks could facilitate the development and exchange of diagnostic protocols, primer and probe sequences, and images of disease symptoms, pathogens, and insect pests and vectors (Tinivella et al. 2008). In addition, the networks could facilitate access to expertise for emerging diseases, pests or pathogens, as well as access to culture collections for reference strains upon which to develop and validate diagnostic protocols.

Under the leadership of the University of Torino, the European Union funded a Crop Biosecurity programme to establish a research network concerned with threatening plant diseases (Gullino et al. 2007, 2008). Among the outputs of the project was the establishment of a list of the most threatening plant diseases, a vulnerability assessment for Europe's agricultural and natural plant systems, a list of diagnostic protocols and technologies, and appropriate mitigation strategies for the most threatening pathogens (Latxague et al. 2007; Gamliel and Fletcher 2008; Suffert et al. 2008). This network should be expanded to include additional European and Mediterranean nations and to incorporate secure communications systems to facilitate collaborative diagnostics (Stack and Baldwin 2008; Fletcher 2008b).

The Global Plant Clinic (http://194.203.77.76/globalplantclinic/) was established to improve the diagnosis of plant diseases in resource-poor nations with poor or non-existent plant diagnostic infrastructure (Boa 2007). The response to the open clinics within rural communities has been tremendous. With little technology, many diseases are being diagnosed and local farmers are being trained to identify common diseases. This model has great utility in serving subsistence farming enterprises globally.

An International Plant Diagnostic Network, funded by the US Agency for International Development, was established involving several US institutions and international partner organisations and institutions (http://www.intpdn.org/). Based on the NPDN model, regional laboratories were established in East and West

Africa, the Caribbean, Central America, and Southeast Asia. Diagnostic training programmes were developed and delivered to the local plant diagnostic clinics in several countries. With sustained funding, IPDN will have significant impact in resource poor nations.

As the benefits of diagnostic networks increase and become more widely known, additional networks will be developed and implemented. Greater co-ordination among these networks may yield a synergy with greater impact in maintaining plant health at the local level while achieving plant biosecurity at the global scale. The challenges to food security and plant biosecurity are increasing while the margins for error are decreasing. The rationale for international co-operation for the diagnosis and management of plant diseases and insect pests is clear; plant biosecurity is prerequisite to food security and public health. This is a global challenge requiring global co-operation; we are all in this together.

References

Audibert M, Mathonnat J, Henry MC (2003) Malaria and property accumulation in rice production systems in the savannah zone of Cote d'Ivoire. Trop Med Int Health 8:471–483

Anonymous (2008) The millennium development goals report 2008. United Nations Department of Economic and Social Affairs, New York, p 56

Boa E (2007) Plant healthcare for poor farmers: an introduction to the work of the global plant clinic. APSnet Feature Story, October 2007. Online. http://apsnet.org/online/feature/clinic/

Brasier CM (2008) The biosecurity threat to the UK and global environment from international trade in plants. Plant Pathol 57:792–808

Brasier CM (2001) Rapid evolution of introduced plant pathogens via interspecific hybridization. BioScience 51(2):123–133

Britton KO (2004) Controlling biological pollution. In: Britton K (ed) Biological pollution, an emerging global menace. APS Press, St. Paul, MN, pp 1–7

Bullock R, Shiel P, Berger P, Kaplan D, Parra G, Li W, Levy L, Keller J, Reddy M, Sharma N, Dennis M, Stack J, Pierzynski J, O'Mara J, Webb C, McKemy J, Palm M (2006) Evaluation of detection techniques to determine presence of Phytophthora ramorum in a nursery. Online. Plant Health Prog. doi:10.1094/PHP-2006-1016-01-RS

Campbell FT (2001) The science of risk assessment for phytosanitary regulation and the impact of changing trade regulations. BioScience 51(2):148

Cochrane H, Haslett D (2002) Deliberate release – what are the risks? N Z J Forest 47:16–17

Collier P (2008) The politics of hunger: How illusion and greed fan the food crisis. Foreign Aff 87(6):67–79

Cupp OS, Walker DE, Hillison J (2004) Agroterrorism in the US: key security challenge for the 21st century. Biosecur Bioterror 2:97–105

Diamond J (2005) Collapse: how societies choose to fail or succeed. Penguin Books, New York, 575 pp

Diamond J (1997) Guns, germs, and steel: the fates of human societies. W.W. Norton, New York, 494 pp

FAO (2001) The state of food and agriculture. FAO, Rome, 289 pp. FAO Agriculture Series No. 33. ISSN 0081-4539

Fletcher J (2008a) The need for forensic tools in a balanced national agricultural security programme. In: Gullino ML, Fletcher J, Gamliel A, Stack JP (eds) Crop biosecurity: assuring our global food supply. Springer, The Netherlands, pp 91–99

Fletcher J (2008b) The case for international cooperation as a strategy to achieve crop biosecurity. In: Gullino ML, Fletcher J, Gamliel A, Stack JP (eds) Crop biosecurity: assuring our global food supply. Springer, The Netherlands, pp 108–118

Fletcher J, Bender C, Budowle B, Cobb WT, Gold SE, Ishimaru CA, Luster D, Melcher U, Murch R, Scherm H, Seem RC, Sherwood JL, Sobral BW, Tolin SA (2006) Plant pathogen forensics: capabilities, needs, and recommendations. Microbiol Mol Biol Rev 70:450–471

Gamliel A, Fletcher J (2008) Crop biosecurity: containment and eradication of invasive pathogens. In: Gullino ML, Fletcher J, Gamliel A, Stack JP (eds) Crop biosecurity: assuring our global food supply. Springer, The Netherlands, pp 71–90

Gullino ML, Fletcher J, Gamliel A, Stack J (2008) Crop biosecurity: assuring our global food supply. Springer, The Netherlands, 148 pp

Gullino ML, Suffert F, Dehne H, Thomas J, Barker I, Gamliel A, Bonifert M, Stack J, Fletcher J, Abd-Elsalam K (2007) Crop and food biosecurity: first results of European research. Phytopathology 97:S44

Lamour KH, Habera LF, Snover-Clift KL, Stack JP, Pierzynski J, Hammerschmidt R, Jacobs JL, Byrne JM, Harmon PF, Vitoreli AM, Wisler GC, Harmon CL, Levy LE, Zeller KA, Stone CL, Luster DG, Frederick RD (2006) Early detection of Asian soybean rust using PCR. Plant Health Prog Online. http://www.plantmanagementnetwork.org/sub/php/research/2006/pcr/. doi:10.1094/PHP-2006-0524-01-RS

Latxague E, Sache I, Pinon J, Andrivon D, Barbier M, Suffert F (2007) A methodology for assessing the risk posed by the deliberate and harmful use of plant pathogens in Europe. EPPO Bull 37:427–435

Mack RN, Lonsdale WM (2001) Humans as global plant dispersers: getting more than we bargained for. BioScience 51(2):95–102

Madden LV, Wheelis M (2003) The threat of plant pathogens as weapons against US crops. Annu Rev Phytopathol 41:155–176

Main CE, Keever T, Holmes GJ, Davis J (2001) Forecasting long-range transport of downy mildew spores and plant disease epidemics. APSnet Feature Story, 22 May 2001. http://www.apsnet.org/online/feature/forecast/

Man in't Veldt WA, Veenbaas-Rijks WJ, Ilieva E, de Cock AWAM, Bonants PJM, Pieters R (1998) Natural hybrids of *Phytophthora nicotianae* and *Phytophthora cactorum* demonstrated by isozyme analysis and random amplified polymorpic DNA. Phytopathology 88:922–929

Palm M (1999) Mycology and world trade: a view from the front line. Mycologia 91:1–12

Rossman A (2001) A special issue on global movement of invasive plants and fungi. BioScience 51(2):93–94

Shelburne EC (2008) The great disruption. The Atlantic September:28–29

Stack JP (2008) Challenges to crop biosecurity. In: Gullino ML, Fletcher J, Gamliel A, Stack JP (eds) Crop biosecurity: assuring our global food supply. Springer, The Netherlands, pp 15–23

Stack JP, Baldwin W (2008) The need for secure communications networks and global connectivity. In: Gullino ML, Fletcher J, Gamliel A, Stack JP (eds) Crop biosecurity: assuring our global food supply. Springer, The Netherlands, pp 103–109

Stack JP, Fletcher J (2007) Plant biosecurity infrastructure for disease surveillance and diagnostics. In: Institute of Medicine (eds) Global infectious disease surveillance and detection: assessing the challenges – finding the solutions. The National Academies Press, Washington, DC, pp 95–106

Stack J, Cardwell K, Hammerschmidt R, Byrne J, Loria R, Snover-Clift K, Baldwin W, Wisler G, Beck H, Bostock R, Thomas C, Luke E (2006) The national plant diagnostic network. Plant Dis 90:128–136

Stack JP, Suffert F, and Gullino ML (2009). Bioterrorism: a threat to plant biosecurity? In, R.N. Strange and ML Gullino (eds) The Role of Plant Pathology in Food Safety and Food Security (In Press)

Strange R, Scott P (2005) Plant disease: a threat to global food security. Annu Rev Phytopathol 2005(43):83–116

Suffert F, Barbier M, Sache I, Latxague E (2008) Biosécurité des cultures et agroterrorisme : une menace, des questions scientifiques et une réelle opportunité de réactiver un dispositif d'épidémiovigilance. Le Courrier de l'Environnement (in press)

Tinivella F, Gullino ML, Stack JP (2008) The need for diagnostic tools and infrastructure. In: Gullino ML, Fletcher J, Gamliel A, Stack JP (eds) Crop biosecurity: assuring our global food supply. Springer, The Netherlands, pp 63–71

Waage JK, Mumford JD (2007) Agricultural biosecurity. Philos Trans Roy Soc Lond B 363: 863–876

Wheelis M, Yokoyama V, Ramos C (2008) Agricultural warfare and bioterrorism using invasive species. In: Heather NW, Hallman GJ (eds) Pest management and phytosanitary trade barriers. CABI, Wallingford, UK, pp 14–19

Whitby S (2002) Biological warfare against crops. Palgrave, Basingstoke

Young JM, Allen C, Coutinho T, Denny T, Elphinstone J, Fegan M, Gillings M, Gottwald TR, Graham JH, Iacobellis NS, Janse JD, Jacques MA, Lopez MM, Morris CE, Parkinson N, Prior P, Pruvost O, Rodrigues NJ, Scortichini M, Takikawa Y, Upper CD (2008) Plant-pathogenic bacteria as biological weapons – real threats? Phytopathology 98(10):1060–1065

Chapter 7
Plant Clinics and Phytopathology Training

H. Maraite, C. Bragard, and A. Legrève

7.1 Introduction

The importance of agriculture has been underlined by the food crisis and the increasing need for renewable energy. Greater efficiency in the use of natural resources, improved sustainability of resources, the adaptation of crops and cropping methods to meet consumer demand and the globalisation of the markets all make the continuous evolution of agriculture imperative.

In recent decades, however, there has been an overall decline in the number of students registering for agricultural degrees at Master of Science (MSc) level in several European countries, as well as a decrease in the number of farmers. Public sector extension services have also declined and agricultural research funding has shifted from long-term projects to short-term complementary funding through competitive grants. Thus, there appears to be a discrepancy between the perceived needs of agriculture and public investment in its development.

This applies to crop protection and, in particular, to phytopathology. There is a great need for more research and more assistance to farmers in view of the many changes in cropping systems (e.g. pathosystems, changes in cultivars, the rapid evolution in pesticide action and resistance, the need to justify pesticide input, the application of pesticide legislation, new integrated pest and crop management concepts, and production constraints imposed by consumer demand for organic agricultural products). Despite this, it is evident that phytopathological research is being carried out within other disciplines (e.g. biotechnology, microbiology, agrometeorology and biochemistry) and by teams with no direct contact with agriculture. The degree to which the outputs of this research are providing solutions to problems is insufficiently assessed and the link to practical applications of the research is often absent.

H. Maraite (✉), C. Bragard, and A. Legrève
Unité de Phytopathologie, Université catholique de Louvain, Croix du Sud, 2/3,
B-1348 Louvain-la-Neuve, Belgium
e-mail: henri.maraite@uclouvain.be

N.V. Hardwick and M.L. Gullino (eds.), *Knowledge and Technology Transfer for Plant Pathology*, Plant Pathology in the 21st Century 4,
DOI 10.1007/978-1-4020-8934-3_7, © Springer Science+Business Media B.V. 2010

There is therefore a need and, to some extent a job opportunity, for graduates capable to combine a knowledge of various fundamental biological disciplines with applied ones, particularly in the complex area of plant health management. Building the capacity of students in plant disease diagnosis and in formulating crop protection advice, appropriately adapted to specific situations, is of prime importance.

In his review entitled 'The role of plant clinics in disease diagnosis and education: A North American perspective', Barnes (1994) highlighted the need for more applied training in plant pathology and noted the significant contribution that plant disease clinics make in this regard. From their surveys of North American universities and plant disease clinics, Aycock (1976) and Evans-Ruhl (1982) reported great diversity in the links between graduate training in plant disease diagnosis and plant clinics, either in house in the case of State land-grant universities, or externally in the case of State departments of agriculture or private services. In European universities involved in agricultural and plant protection training at MSc level, the extension-research-academic training link is generally not as strong as it is in the American land-grant universities. Several plant pathology or plant health units or departments provide a diagnosis and advice service on a voluntary basis, often without a specifically defined structure. This makes it difficult to use a simple internet survey to assess the potential of integrating plant clinics into phytopathology training at European universities. In preparing the contribution for ICPP2008, we had initially planned a detailed survey, but from our experience with various partners in the Erasmus-Socrates programme and a rapid analysis of phytopathology curricula featured on the internet we realised that the situation was also quite heterogeneous in Europe. It would therefore be difficult to reach clear conclusions. Each university has its own economic environment, its own capacities and constraints, and therefore may have different approaches towards reaching the same training goal. In the absence of a performance evaluation system, it is not advisable to draw conclusions about the relative value of the various approaches.

We therefore decided to restrict this presentation to the experience we have gained and to use it as a basis for the exchange of ideas. We shall outline the agricultural context and agricultural research, extension and education at MSc level in Belgium and focus on the development of phytopathology training linked to the Plant Clinic at UCL.

7.2 The Belgian Agricultural and Educational Context

Belgium is a small country, 30,528 km² in area (1/18 the size of France) with 10.1 million inhabitants. There are three language Communities (Dutch, French and some German) and three regional entities (Flanders, Wallonia and Brussels). Higher education in the country reflects these regional differences, particularly in agronomy, and adds another layer of fragmentation: Community-organised universities in Gent (Universiteit Gent) and Liège-Gembloux (Faculté Universitaire des Sciences Agronomiques de Gembloux, FSAGx), catholic universities in Leuven

(Katholieke Universiteit Leuven, KUL) and Louvain-la-Neuve (Université catholique de Louvain, UCL), non-denominational universities in Brussels (Université Libre de Bruxelles, ULB) and Antwerp (Vrije Universiteit Brussel – Universiteit Antwerpen, VUB-UA), making a total of six institutions delivering an MSc degree in agronomy, referred to as 'bioengineering'.

The job market is also quite diverse. Flanders is characterised mainly by intensive outdoor and protected vegetable, fruit and ornamental production, whereas in Wallonia wheat, barley, potato and sugar beet crops, as well as pasture and silage maize-based milk and meat production from cattle, are dominant.

Agricultural production policy and support have been regionalised for about a decade. Agricultural research is supported at regional level through competitive grants for the universities and higher educational institutions and for the regionally funded agricultural research centres in Merelbeke and Gembloux, as well as for crop-specific research and experimental institutions or stations partially supported by growers.

Public investment in extension has fallen dramatically in the past 30 years. Most regionally paid agricultural support staff at Bachelor of Science (BSc) and MSc level who are in contact with farmers are involved in guiding farmers through the ever-increasing administrative requirements or in ensuring their compliance with legislative constraints.

Extension has been handed over to technical crop-specific institutions and to non-profit associations partially supported by regional authorities; some of them are linked to university departments. Technicians in private companies supplying seeds, fertilisers and pesticides to the farmers also provide technical guidance on crop protection. This differs from the approach used in the USA for plant pathology extension services (Jacobsen and Paulus 1990).

Although the national and regional authorities have withdrawn from carrying direct responsibility for agricultural extension, the continuous and sometimes radical changes in Belgian agriculture continue in response to changes in European Union (EU) agricultural policy and to the need to keep pace with technology. Changes in cropping systems, the increased exchange in agricultural products and the effect of climate change have all brought about changes in pathosystems. The decline in the percentage of Belgium's population involved in agricultural production, however, and the increased use of pesticides has triggered public distrust in agriculture. It has also resulted in the falling numbers of students interested in studying agricultural subjects at university, despite the fact that the need for agricultural expertise is greater than before, especially with regard to crop protection.

7.3 Agricultural and Crop Protection Training at UCL

The declining interest in training in agronomy led to change in the MSc title in Belgium, from 'agricultural engineer' to 'bioengineer', and a change in the name of our faculty from Faculty of Agronomical Sciences to Faculty of Biological,

Agronomical and Environmental Engineering. Training programmes have also undergone changes and there are now three MSc programmes in bioengineering: Agronomy; Chemistry and Bio-industries; and Environmental Sciences and Technologies. Each programme offers students various options depending on their training objectives. Our training of students in the MSc in 'Bioengineer Agronomy', and in particular the Integrated Crop Protection (BIR239A) option, takes account of the needs of Belgian agriculture and the fragmented job market, with few opportunities for some specialisations.

We offer all 'Bioengineer Agronomy' students a solid foundation in cropping and husbandry systems, data and process management, biotechnology, economics and an introduction to crop protection (Principles of Phytiatry). In the Integrated Crop Protection option, specific courses providing advanced knowledge in crop protection account for almost half the 120 ECTS required (European Credit Transfer and Accumulation System; 1 ECTS stands for a workload of about 25–30 working hours) (http://ec.europa.eu./education/ programmes/socrates/ects/index_en.html#1).

Advanced training in crop protection, without affecting the need for multidisciplinarity and for adapting to changes in agriculture and the job market, has been the main training objective since the creation of the Integrated Crop Protection option at our Faculty in 1967 and during the several programme revisions since then. Apart from the research-based MSc thesis with a credit of 27 ECTS, the compulsory disciplinary training in crop protection includes courses on: factors affecting plant development, with lectures on viruses, bacteria, fungi, nematodes, arthropods and weeds; case studies of negative pathogen or pest interactions with plants in the phytopathology and applied entomology courses; pathogen-plant environment interactions in epidemiology; and crop protection strategies in the courses on crop improvement, phytopharmacy, biological and integrated control. The capacity to integrate knowledge across disciplines is strengthened by such courses as Recent Advances in Crop Protection (3 ECTS) and Plant Clinic (6 ECTS). UCL is also linked to the international supplementary MSc in Protection of Tropical and Subtropical Crops (TROPMC2), co-organised with FSAGx and the Centre de Coopération Internationale en Recherche Agronomique pour le Développement – Ecole Nationale Supérieure Agronomique de Montpellier (CIRAD-ENSAM), France.

7.4 The Plant Clinic Course at UCL

The Plant Clinic course has been a cornerstone of our crop protection curriculum at UCL since 1976. The introduction of this course stemmed from the author's experience of the plant clinic linked to the Department of Botany and Plant Pathology at Purdue University, Indiana, USA, during a postdoctoral visit there in 1972. Under the supervision of senior staff, MSc students in plant pathology at Purdue University were responsible for analysing diseased plant samples brought in by extension officers.

Plant clinic training appeared to be an effective approach for consolidating the phytopathology training by enabling students to apply the various diagnostic techniques and concepts learned during their MSc studies to often untypical situations in the field.

At that time, we were questioning UCL's approach towards practical training in plant pathology. In reaction to earlier training in identifying fungal diseases based on herbarium samples and microscopic slides, and to improve the understanding of the mechanisms of pathogenesis and, therefore, of the value of integrated control measures, we had moved in 1968 to plant pathology exercises based, *inter alia*, on the useful 'Sourcebook of Laboratory Exercises in Plant Pathology' (Kelman et al. 1967). Students were advised to learn during the summer months to identify the most important diseases, referring to the numerous field guides and illustrations available.

We soon realised that this was wishful thinking and that the students' disease recognition capacity had seriously decreased. It was also clear, however, that identifying the cause of a disease and understanding the conditions leading to yield loss were not sufficient; a farmer also expects advice on how to control or prevent the disease in his/her particular situation. We therefore decided to reintroduce practical training in disease diagnosis combined with the formulation of crop protection advice.

The application of the plant clinic concept, as implemented in an American land-grant university, to the UCL situation did not seem feasible, as funding a plant clinic service was not a UCL priority. Students were carrying a heavy course load in the context of a multidisciplinary approach, and meeting these needs was incompatible with the quality service that we wanted to offer clients. We therefore established a Plant Clinic course. In line with Aycock (1976) and Evans-Ruhl (1982), we see a 'plant (disease) clinic' as offering 'plant doctor' training, using actual specimens of diseased plants from the field and applying integrated scientific knowledge to the understanding, prevention or control of the observed disease. The main goal of a plant clinic is to increase the competence of graduates when faced in their professional lives with plant diseases, whatever the situation. Beyond being familiar with plant health practices, their responsibility as plant doctors and the challenges of keeping abreast of scientific progress in their professional lives is important.

A regular interactive group-learning process on a range of problems is essential to achieve this goal. The pre-requisites for the learning process are the full assimilation of the courses on plant pathogens and pests and on phytopathology and applied entomology. The Plant Clinic course at UCL lasts for the whole academic year (30 weeks) on a half a day/week basis for the BIR239A and the SOCRATES or TROP2MC students.

Because of time constraints the Plant Clinic course starts at the same time as the courses on crop protection strategies, whose elements are progressively discussed in the process of formulating practical advice for solving the analysed disease problem. Group size is critical for fruitful interaction. The optimum is between five and ten students. It is also important that the students have a similar background and are at the level of the majority of the group or are able to reach that level quickly by themselves. When there are more than 15 registered students, they are split into groups of 5–10.

Supervision is provided by two professors of phytopathology, with one assistant and technical support.

7.4.1 Content of the Plant Clinic Course

The Plant Clinic course starts in the autumn with one or two outside sessions aimed at training the students to recognise disease or pest injury symptoms in situ, analyse the distribution of the symptoms, assess disease incidence and severity, and evaluate the risk of further disease development. The sessions involve on-the-spot discussions on possible causes, checking for predisposing factors in the environment, sampling and devising conservation strategies for further diagnosis in the laboratory.

The supervisors usually trigger the process by drawing attention to diseases that might have been overlooked. The students are then invited to detect others and the results are discussed.

During the following 10–15 weeks, the students are trained to become more efficient in disease diagnosis in the laboratory. The laboratory for the Plant Clinic course has 15 student workstations equipped with a Leica M3/M5 stereomicroscope, a Zeiss Axiomat light microscope with phase contrast and dark field, a box with tools and reagents for slide preparations, and isolations of pathogenic fungi and bacteria. There is also a microscope and a stereomicroscope equipped with a video camera for demonstrating various isolation procedures and providing guidance on spotting key features in microscopic slides. A computer with an internet connexion allows students to search databases rapidly. This is all backed up by collections of preserved disease samples, slides and basic literature, such as the American Phytopathological Society compendium series, International Mycological Institute (IMI) descriptions of fungi and bacteria, and a collection of various reference books (e.g. Anon 1999; Alford 1994; Barnett and Hunter 1998; Bergmann 1992; Brandenburger 1985; Ellis and Ellis 1997).

Hypotheses about possible causal agents are formulated by comparing the observed symptoms with those in pictures or in descriptions, and the approaches for testing these hypotheses are discussed. The students' attention is drawn to the difficulties and pitfalls of diagnosis based on pictures alone.

During the spring, more field trips and visits to other diagnosis laboratories are organised in order to give students experience in working in the field and to broaden their ability to set up diagnosis facilities.

7.4.1.1 Fungal Diseases

For the diagnosis of fungal diseases considerable effort is made to optimise the preparation and examination of microscopic slides from diseased plant samples. The options for the various phases are discussed step by step, encouraging input from the students and comparing the results. Lack of proficiency in this basic but key operation often causes a bottleneck because it requires complete knowledge of the organism being examined, as well as the level of intuition acquired with practice. Starting with the same disease for all the students, the disease spectrum is progressively diversified, with the students writing their observations on the blackboard and sharing them in group discussions.

The next step involves comparing the observations under the microscope with descriptions in the available literature. The students gradually learn which sources to consult for a given problem.

A similar approach is used for the isolation and culture of potential fungal pathogens. Discussions focus on the isolation media to be used, their preparation and possible amendment for increased selectiveness, sample preparation and isolation methods, incubation conditions for growth and/or sporulation, and preparation and conservation of single spores or contaminant-free cultures.

The identification of some cultures by electron microscopy, or polymerase chain reaction (PCR), is demonstrated in collaboration with the mycological reference laboratory, Mycothèque de l'UCL, a member of the Belgian Coordinated Collections of Microorganisms (MUCL/BCCM) housed in the same building. The use of Koch's postulate is discussed for particular diseases. Some weeks are devoted to using standardised in vitro fungicide sensitivity tests developed by our group.

Apart from these classical methods, small teams of students are also trained in the practical use of ELISA tests and PCR tests for mycotoxin and fungal detection, as well as in biotests for assessing the inoculum potential in soil.

During the course, more attention is gradually given to formulating disease control advice adapted to the particular situation in which the disease is developing. Students are encouraged to integrate knowledge gained from their course with a critical review of disease control measures reported in the literature. Particular attention is given to an integrated approach and to match the possible recommendation of pesticides use with the fast-evolving regulations and lists of authorised products (http://www.fytoweb.fgov.be/indexen.htm).

7.4.1.2 Bacterial Diseases

Real problems from the field (e.g. *Xanthomonas* on grains and grasses, *Pectobacterium* on potato, *Pseudomonas* on beans, and apple proliferation) are analysed, highlighting the importance of going beyond symptom-based diagnosis of bacterial diseases by a simple examination under the stereomicroscope, the dark field or the phase contrast microscope.

The students are also trained in the isolation of bacteria, following group discussions on where and how to isolate, on what media to isolate, and after which treatment. Isolates are then characterised on the basis of colony appearance, Gram staining and various physiological, biochemical, hypersensitivity and pathogenicity tests performed by each student and the results are compared and discussed.

Under the guidance of a supervisor, students assess the practical value of various detection and identification tests, including serology, immunofluorescence, dot-blot, antibody production and pros and cons of polyclonal antibodies

(Pab), monoclonal antibodies (Mab), sodium dodecyl sulfate polyacrylamide gel electrophoresis (SDS-PAGE), fatty acid methyl ester profile (FAME), PCR, restriction fragment length polymorphism (RFLP), amplified fragment length polymorphism (AFLP), microarray, and sequencing of ribosomal deoxyribonucleic acid (rDNA). They are also introduced to the use of references for identifying plant pathogenic bacteria.

7.4.1.3 Virus and 'Virus-Like' Diseases

The field problem-based approach (e.g. rhizomania, peanut clump, potato virus Y, pepino mosaic, and viroids), together with documented case studies or scenarios, is used. Questions are addressed in turn to individual students or teams of two or three students in order to stimulate knowledge sharing among them. The training comprises:

- Systematisation of symptom-based diagnosis of viral diseases, light microscope detection of cellular inclusions, sample preparation and handling the international exchange of samples
- Familiarisation with virus databases (International Committee on Taxonomy of Viruses (ICTV), http://www.ncbi.nlm.nih.gov/ICTVdb/; Description of Plant Viruses (DPV), http://www.dpvweb.net/; National Centre for Biotechnology Information (NCBI), http://www.ncbi.nlm.nih.gov/genomes/genlist.cgi?taxid=10239&type=5&name=Viruses)
- Virus indexing exercises after discussions on the choice of host and method
- Demonstration of the use of transmission electron microscopy (TEM) for dip analysis, and discussions on more advanced techniques, such as immunosorbent electron microscopy (ISEM), decoration, and immunogold labelling
- Familiarisation with serological techniques, such as enzyme-linked immunosorbent assay (ELISA), immuno-diffusion and Western blot, as well as reverse transcription polymerase chain reaction (RT-PCR)/PCR-based techniques and virus genome analysis

7.4.1.4 Herbarium

In order to encourage self-training in diagnosis and the formulation of disease control advice adapted to specific cropping situations, students also have to create a herbarium of 20 personally collected and identified samples of diseases (on any crop) caused by fungi, bacteria or pests. The diseases must differ from those collectively analysed during the Plant Clinic course and the phytopathology exercises. Apart from a sampling sheet specifying the conditions under which the disease had occurred, the students have to provide all the information to support their diagnosis (e.g. drawings from structures seen under the microscope with the microscopic slide), as well as disease control advice and the references used. The herbarium is assessed for the originality of the collected samples, accuracy of diagnosis and pertinence of the advice.

7.4.2 Plant Clinic Capacity Testing

The students' ability in diagnosis and formulating advice is evaluated at the examination of diseases not analysed during the courses. For fungal diseases, the examination is made up of samples to be analysed under the microscope; for bacterial and viral diseases the examination is based mostly on pictures of symptoms, together with details of the conditions in which the disease occurred.

After 30–45 min, without referring to notes or documentation, the students have to deliver a sheet with the name of the suspected causal agent and the reasoning behind their diagnosis, their advice to the farmer on how to restrict or control the disease, what further tests should be performed in the laboratory and a short list of references to consult.

This is followed by a period of 65–90 min during which the students have access to their notes and to the library in order to consolidate the diagnosis, as well as access to the website of the Belgian Federal Public Service Health, Food Chain Safety and Environment (http://www.fytoweb.fgov.be) to check on authorised plant protection products. For bacterial and virus diseases, a complete analysis is simulated with the supervisor, the student proposing a test, the supervisor giving possible results of the test. Here, good methodology is considered as important as the accuracy of the diagnosis. Students failing in the diagnosis are nevertheless assessed for their formulation of advice on disease control in the situation described. Because of the strict time limit, the students have to draw on knowledge acquired during the whole course, rather than relying only on their notes and the reference material.

The phytopathology background of students registered for the TROP2MC programme is more heterogeneous than those on the regular BIR239A programme who will have done most of their courses at UCL. In some cases, a lack of practical skills in the preparation and analysis of microscopic slides and in identifying phytopathogenic fungi hampered the creation of a satisfactory herbarium (Fig. 7.1). However, through the complementary training in crop protection and intensive practicals during the plant clinic course, most of them achieved a similar level of attainment in the examination on fungal disease as the BIR239A students had. A link between the degree of attention to the herbarium preparation and proficiency in identifying a new unknown sample was evident among the TROP2MC students, but not among the BIR239A ones, whose marks also tended to be less variable.

The final mark given by the Plant Clinic for fungal diseases is the mean of the marks for the herbarium and the examination. From 1999 to 2008 these marks for fungal diseases were similar to those for the examination on virus and bacterial

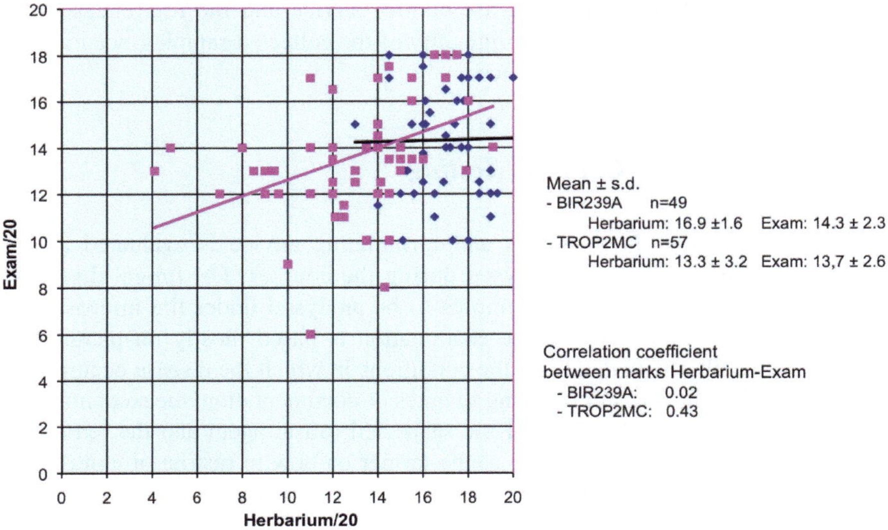

Fig. 7.1 Relationship between the herbarium and examination marks in the Plant Clinic course at UCL, in the fungal diseases section, for students working for an MSc in Bioengineer Agronomy, option Integrated Crop Protection (BIR239A, *diamonds*) and for students working for the international supplementary MSc in Protection of Tropical and Subtropical Crops (TROPMC2, *squares*)

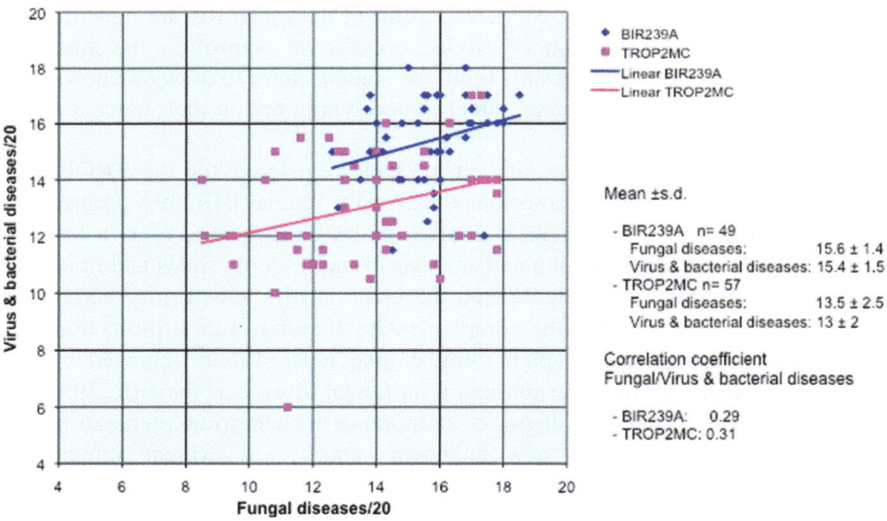

Fig. 7.2 Correlation between the marks for the sections on fungal diseases and on virus and bacterial diseases in the Plant Clinic course at UCL for students working for an MSc in Bioengineer Agronomy, option Integrated Crop Protection (BIR239A) and for students working for the international supplementary MSc in Protection of Tropical and Subtropical Crops (TROPMC2)

diseases (Fig. 7.2), although there was some variation in individual marks. The overall means for the TROP2MC students were lower than for the BIR239A students, due to the difficulties some of them had in catching up with the group.

7.4.3 *Assessment of the Plant Clinic Course*

During the 30 years or more that we have been organising the Plant Clinic course at UCL, the types of diseases, the techniques and the modes of organisation have, of course, evolved, but the basic principles have stood up to periodic evaluations and have been kept.

The course is very demanding for the students and the supervisors, but the outcome has always been rewarding. The moment when students master producing an independent diagnosis varies from year to year, depending on the groups and on individual students, but it generally occurs after about 22 weeks. Delays in reaching this goal can be frustrating and usually encourage extra personal investment. Failures to reach the goal are rare, especially for the BIR239A students. However, students who have not received enough training in plant pathology and plant pathogens during their ERASMUS-SOCRATES period, or have come from other universities, such as the overseas TROP2MC students, may struggle a little to catch up.

The problem-based approaches, the active involvement of students through discussions and practice over 30 weeks greatly encourage their interest in crop protection challenges and, in particular, in phytopathology. This accords with the analysis of inquiry-based learning by Schumann (2003) in his review of innovations in teaching plant pathology. By acquiring a solid understanding of the discipline and its integration into multidisciplinary-based bioengineering training, students are well prepared for their future careers, even if few will actually work as plant doctors.

A comparison of the marks gained by 45 students for the introductory course, Principles of Phytiatry, and for the Plant Clinic course showed an increase in the mean from 12.7/20 to 15/20 (Fig. 7.3). The proportion of poor marks (<12/20) fell from 29% to 2%, while the good marks (≥14/20) rose from 31% to 89%. This indicated that the overall proficiency of the group increased with a slight narrowing of the difference among students.

7.5 The Plant Clinic Service at Louvain-La-Neuve

In addition to the Plant Clinic course, the professors and scientists at our Phytopathology Laboratory at UCL were regularly approached for diagnosis and advice on diseases in a wide range of European or overseas crops. This led to the identification of new research topics and a good reputation for our group. Nevertheless, as demand increased, the Plant Clinic work began to interfere too much with ongoing activities.

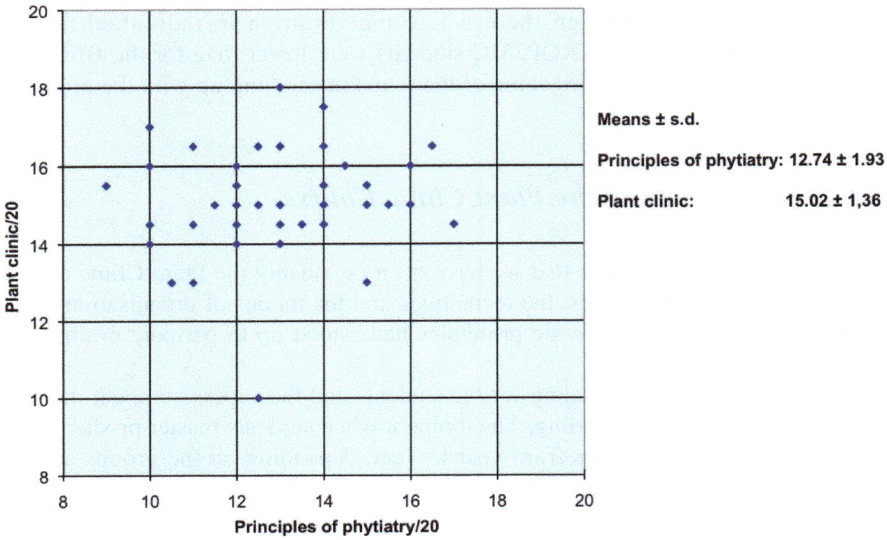

Fig. 7.3 Comparison of the marks for the introductory course, Principles of Phytiatry, with those for Plant Clinic at UCL for BIR239A students from 1997 to 2006

Taking into account the importance for our group to have an interface between research-teaching and extension, and also the administrative and financial difficulty of UCL becoming involved in extension work, we have set up a non-profit association with colleagues from the university and others interested in agricultural development. We have called this association Co-ordination of Research and Rural Development (CORDER).

Since 1985, under the umbrella of CORDER we have developed the Plant Clinic service. Initially, staff were paid through a regional unemployment-reduction programme supporting activities of public interest. Over the years the budget has increased to about €350,000/year and is covered by the clients paying for services, by service agreements with public and private organisations and in some years by an allowance from the Walloon Region for the public service offered.

The Plant Clinic is staffed by two MSc graduates in crop protection, one MSc level bioengineer to help implement the quality management programme, and the equivalent of three and a half part-time technician positions. CORDER has an agreement with UCL for renting space at the UCL Laboratory of Phytopathology and for sharing some heavy equipment, and for using the library and the L3 biosafety laboratory for work with quarantine organisms. The team also works with the Phytopathology Laboratory on the development of new diagnostic tools and on identifying fungi in the MUCL/BCCM collection using molecular biology. We collaborate closely with the Comité Regional PHYTO, a platform for dialogue and action on the responsible use of pesticides, set up by the Walloon Region at the UCL Laboratory of Phytopathology.

Apart from typical plant clinic work involving diagnosis and advice, the Plant Clinic team is also a leading partner in crop protection extension activities on cereals, potato and horticultural crops. For these activities the team relies partly on research results from the Phytopathology Laboratory, such as the web-based interactive decision support system, Proculture, for fungicide protection in winter wheat against *Mycosphaerella graminicola*. The Plant Clinic team manages the Proculture website (www.fymy.ucl.ac.be/proculture/) and incorporates the simulations from the system into its weekly forecasting during the critical period of the wheat growing season. There is also a feedback to the scientists on ideas for improvement and on emerging problems.

The Plant Clinic service is therefore truly an interface between the university and the farmers. Its work would not be viable, however, without the income generated by standardised phytosanitary analysis (e.g. of potatoes for the presence of quarantine and soft-rot bacteria, of industrial waste for the absence of plant pathogens, fungicide sensitivities assessments, biotests for assessing the inoculum potential of *Aphanomyces*, serological quantification of mycotoxins, race typing of popular rust races, and so on).

Between 2000 and 2007 the number of samples analysed increased annually from 1,000 to 3,544. From the 369 requests for diagnosis and analysis out of the 3,544 received in 2007, almost half originated from individual farmers or citizens and a quarter from officials (Table 7.1).

Most of the samples were potatoes, in need of analysis for the detection of the quarantine bacteria *Ralstonia solanacearum* and *Clavibacter michiganensis* subsp. *sepedonicus*. A substantial amount of the other samples were from various horticultural crops (Table 7.2). This contrasts with the situation 10 years ago when most of the samples analysed were cereals, highlighting the rapid evolution of the type of analysis requested and the need for flexibility to respond quickly to clients' needs.

The increase in the proficiency of phytosanitary analysis was made possible by the building up of a portfolio of standardised tests and of a quality management system in order to give the quality assurance required by some clients. Quality assurance in plant health diagnostic laboratories is becoming a key issue (Thrane 2008).

In April 2008 we became the first laboratory in Belgium to receive accreditation according to NBN EN ISO/IEC 17025:2005 for the detection of quarantine bacteria on potato.

Table 7.1 Categories of clients who requested diagnosis and phytosanitary analysis from the Louvain-la-Neuve Plant Clinic service in 2007

Clients	Number of requests
Individual farmers or citizens	164
Companies, research centres, laboratories	117
Federal or Regional institutions	88
Total	369

Table 7.2 Type of samples analysed by the Louvain-la-Neuve plant clinic service in 2007

Type	Number of requests	Analysis
Potato	142	3,044
Cereals	20	24
Sugar beet and chicory	7	10
Maize	1	1
Rape seed	1	1
Vegetables	20	51
Fruits	19	22
Trees, shrubs	61	91
Ornamentals	49	105
Turf and silage	5	5
Wastewater	41	180
Various	3	10
Total	369	3,544

The Plant Clinic service was associated with the Phytopathology Laboratory in setting up a plant clinic at Kinshasa as a focal point for other plant clinics in DR Congo, and is also associated with the plant clinic at Hue University, Vietnam.

7.6 Relationship Between the Plant Clinic Service and the Plant Clinic Course

There is a synergy between the non-profit Plant Clinic service and the university-offered Plant Clinic course. Sometimes, interesting samples submitted to and analysed by the service are also analysed during the course. Plant Clinic documentation is shared by the course and the service, and through links with the service the students also become familiar with the functioning and requirements of a Plant Clinic service. At the end of their training, usually in May, the most experienced students are given the opportunity to participate in the analysis of some samples submitted to the Plant Clinic service. After the analysis they submit their diagnosis and recommendations to the service bioengineer. These samples can be included in their Plant Clinic course herbarium. Some students work with the Plant Clinic service for their MSc thesis research. Specific training in techniques used by the Plant Clinic service is also arranged.

Before their graduation, however, MSc students are not formally associated with the Plant Clinic or the phytosanitary analysis work of the service, as is the case in some American land-grant universities. This is because of the difficulty in matching the expectations of the clients for a fast, good quality service with the tuition and guidance that would be needed for the students, which is beyond the duties of the service in our present funding situation. Training an annually changing, and sometimes high, number of students on the Plant Clinic course (29 students in 2008) cannot be integrated into the activity of the service. Confidentiality and quality management requirements, such as the full training of staff before involvement in the analysis, are also major constraints.

7.7 Conclusions

The Plant Clinic course has been valuable in deepening the knowledge of practical crop protection, particularly in terms of phytopathology, in the line with the profession of plant doctor. The plant doctor professional status, however, has not been formalised. The author can recall the enthusiastic debates on the 'Phytomediziner' profession at ICPP78 in Munich. Only a small proportion of our graduates, however, are doing Plant Clinic work as part of their professional career.

Most of the Plant Clinic students appreciate the course and the intensive work and group discussions associated with it. They consider the knowledge integration through group discussion as positive, although the brightest students are sometimes impatient if the discrepancy in the level of pre-course training in the group is too wide. A solution has been found to this by organising two groups. Student guidance and supervision is very demanding. The rewards are satisfaction with the progress observed and the positive feedback from employers of our graduates on their solid practical training in crop protection and their problem-solving capacity.

This fosters our conviction about the value of Plant Clinic training for students graduating in crop protection and in the transfer of the knowledge from researchers to farmers. Training the students in a Plant Clinic context, as done in some American land-grant universities, is a valuable option and we fully agree with Barnes's (1994) statement: 'What more logical a place … to use basic plant pathology concepts to formulate strategies for plant disease control'. In our situation of a university not involved *per se* in extension, however, it seemed better to run the Plant Clinic course parallel to the Plant Clinic service, the latter being a good way for students to see the relevance of their training.

There is no single way for associating plant clinics with training in phytopathology. Each university has to find the way that suits it best, depending on its resources, environment and goals. For the plant doctor profession, however, it is important to keep in mind the need to bridge, through plant clinic training, the ever-increasing translation gap between research and crop protection practice, as highlighted recently for human medicine (Butler 2008).

References

Alford DV (1994) Ravageurs des végétaux d'ornement – arbres, arbustes, fleurs. Institut National de la Recherche Agronomique, Paris, France, 464 pp

Anonymous (1999) Guide pratique de la défense des cultures. Association de Coordination Technique Agricole, Lille, France, 588 pp

Aycock R (1976) The plant disease clinic-a thorn in the flesh, or a challenging responsibility. Annu Rev Phytopathol 14:165–175
Barnes LW (1994) The role of plant clinics in disease diagnosis and education: a North American perspective. Annu Rev Phytopathol 32:601–609
Barnett HL, Hunter BB (1998) Illustrated genera of imperfect fungi. The American Phytopathological Society Press, St. Paul, MN, 218 pp
Bergmann W (1992) Nutritional disorders of plants: development, visual and analytical diagnosis. Gustav Fischer Verlag, Stuttgart, Germany, 741 pp
Brandenburger W (1985) Parasitische Pilze an Gefässpflanzen in Europa. Gustav Fischer Verlag, Stuttgart, Germany, 1248 pp
Butler D (2008) Crossing the valley of death. Nature 453–7197:840–842
Ellis MB, Ellis JP (1997) Microfungi on land plants. An identification handbook. The Richmond Publishing Co., Slough, UK, 868 pp
Evans-Ruhl G (1982) Plant disease clinics – past, present, and future. Plant Dis 66:80–86
Jacobsen BJ, Paulus AO (1990) The changing role of extension plant pathologist. Annu Rev Phytopathol 28:271–294
Kelman A, et al (1967) Sourcebook of laboratory exercises in plant pathology. Sourcebook Committee of the American Phytopathological Society, W.H. Freeman, San Francisco, CA, 387 pp
Schumann GL (2003) Innovations in teaching plant pathology. Annu Rev Phytopathol 41:377–398
Thrane C (2008) Quality assurance in plant health diagnostics-the experience of the Danish Plant Directorate. Eur J Plant Pathol 121:339–346

Chapter 8
Training in Plant Pathology from an Industry Perspective

U. Gisi

8.1 Introduction

Plant pathology is defined as the science of micro-organisms and the environmental factors that cause disease in plants including the mechanisms by which these factors cause diseases (host–pathogen interactions) and the methods of preventing and controlling disease. The causative agents of plant diseases include pathogenic micro-organisms such as viruses, bacteria, fungi and related organisms (oomycetes), protozoa and nematodes, as well as unfavourable environmental conditions (abiotic factors), such as lack or excess of nutrients, moisture, light, temperature and presence of toxic chemicals. In most cases diseases can be recognised easily by the appearance of symptoms. In addition to systematical grouping based on morphology, spore types and molecular relatedness, plant pathogens can be grouped also according to the types of symptoms they cause. Both aspects are important tools for teaching programmes in plant pathology and help to estimate the importance of a given pathogen in a given crop.

University teachers tend to favour diseases and pathogens in their teaching courses about which extensive knowledge of genomic information is available, such as downy mildew (*Hyaloperonospora arabidopsis [= Peronospora parasitica ssp arabidopsis]*) on *Arabidopsis*, rust (*Uromyces appendiculatus*) on bean, powdery mildew (*Blumeria [= Erysiphe] graminis fsp. hordei*) on barley or single pathogens like *Sacharomyces cerevisiae, Aspergillus nidulans, Ashbya gossypii*. On the other hand, applied plant pathologists and industry researchers are often obliged to deal with agronomically important pathosystems and pathogens such as downy

U. Gisi (✉)
Syngenta Crop Protection, Biology Research Center, Schaffhauserstrasse, CH-4332 Stein, Switzerland
and
Institute of Botany, University of Basel, Hebelstrasse, CH-4056 Basel, Switzerland
e-mail: ulrich.gisi@syngenta.com

N.V. Hardwick and M.L. Gullino (eds.), *Knowledge and Technology Transfer for Plant Pathology*, Plant Pathology in the 21st Century 4,
DOI 10.1007/978-1-4020-8934-3_8,

mildews (and *Phytophthora* and *Pythium* spp.) on grapes, potato and vegetables; powdery mildews on cereals, grapes and vegetables; rusts on cereals, soybeans and coffee; leaf spots on a range of crops, especially cereals, rice and many field and tropical crops, as well as single pathogens like *Phytophthora infestans, Mycosphaerella* spp. (especially *M. graminicola*) and *Fusarium* spp. All aspects of plant pathology have to be covered for these pathogens and the diseases they cause during training and teaching programmes by industry researchers, such as disease cycles, pathogenicity and development stages, description of symptoms, spore types and formation, processes of apical growth and spore germination (cytological, biochemical, physiological aspects), different types of interactions between pathogen and host (defence reactions, resistance, pathogenicity, virulence, aggressiveness), epidemiology and population genetics, and last but not least disease control strategies and fungicide resistance.

Since the spectrum of diseases is extremely large and many crops are involved, it is impossible to provide information for all pathosystems. Therefore, the main focus must be given to those diseases and pathogens which cause the most devastating damages of major crops. A simple system was developed defining the importance of pathosystems for teaching undergraduate students at universities but also for designing screening concepts in the discovery process of fungicides in agrochemical companies. With a similar approach, a simple overview is given describing the spectrum of activity of fungicides. For more advanced students (Master and Ph.D. candidates), the understanding of genetics and molecular processes in pathogens is becoming more important and can be explained in a simple way by using fungicide resistance as a marker to follow inheritance in fungi (which are mostly haploid) and related organisms like oomycetes (which are diploid).

8.2 Which Are the Most Important Symptoms of Plant Diseases?

Two major types of compatible interactions (trophies) between pathogens and host plants can be distinguished, that is the biotrophic and perthotrophic (or necrotrophic) types. Examples of biotrophic interactions with the formation of the typical haustoria as nutritional bodies are downy mildews (oomycetes), powdery mildews (ascomycetes) and rusts (basidiomycetes), whereas fruit rot pathogens, moulds, leaf spot and soil-borne pathogens are necrotrophic organisms producing typical disease symptoms (Table 8.1). Some of the latter organisms can also colonise and degrade dead tissue thus representing saprotrophic life stages. Seed-borne pathogens can cover biotrophic and perthotrophic stages and sometimes express their symptoms during flowering of the host plant only after one plant generation (smuts and bunts).

Table 8.1 Important symptoms of plant diseases and causal agents

1. Visible (white, grayish, light brown) spore mat (sporangiophores and sporangia) mostly on lower leaf surface (biotrophic pathogens)
Downy mildews (and *Phytophthora* spp.) on leaves, fruits, shoots
2. Visible white mycelial and powdery spore mat or pustules (biotrophic pathogens)
Powdery mildews on leaves, fruits, shoots
3. Visible red-brownish spores and pustules (biotrophic pathogens)
Rusts on leaves, stems
4. Fruit rots and moulds (perthotrophic/saprotrophic pathogens)
Stems, roots, leaves, fruits (*Monilinia, Botrytis, Glomerella, Penicillium,* etc.)
5. Necrotic leaf spots and die-back, defoliation (perthotrophic pathogens)
Leaves, stems (*Septoria, Alternaria, Cercospora, Pyricularia, Venturia,* etc.)
6. Damping-off and wilting (soil-borne pathogens)
Stem base and root diseases, rot (*Pythium, Rhizoctonia, Oculimacula,* etc.)
Blockage of vascular system, toxins (*Verticillium, Fusarium,* etc.)
7. Systemic plant and seed diseases (seed borne pathogens)
Smuts and bunts
8. Bacterial (and viral) symptoms like die-back, rot, leaf spots, discolouration

Not included in this list are plant defence reactions such as the formation of cork (scab), malformation of organs (canker, tumours, curling, hypertrophy), gum secretion and hypersensitive reaction (spots).

8.3 Which Are the Most Important Diseases on Important Crops?

Instead of grouping diseases according to the systematic position of the causal pathogens in the kingdom of organisms (Oomycetes, Chytridiomycetes, Zygomycetes, Ascomycetes including Deuteromycetes, Basidiomycetes), typical disease symptoms (Table 8.1) have been used to classify the respective pathogens. Seven types of symptoms induced by fungal plant pathogens, plus one caused by bacteria (and viruses) (Table 8.1), represent the eight columns in the Disease – Crop Matrix which includes 14 key crops (lines A to O in the Disease – Crop Matrix) (Fig. 8.1a). By combination of letters and figures (e.g. A3), the corresponding pathogen name (pathogen genus, e.g. *Puccinia* for cereal rust, A3) can be defined (Index, Fig. 8.1b). The crops are sorted according to the market size of the products which are used to control the corresponding disease. The extent of damage produced in the single crop – disease systems is two plus (++) for major loss, one plus (+) for minor loss and blank for not existing or economically not important.

In this matrix, a range of crop – disease systems are especially important, such as leaf spot diseases on all crops; fruit rots in grapes, pome and stone fruits, tropical crops and vegetables; downy and powdery mildews in grapes, potato, vegetables, turf and ornamentals; rusts on cereals and soybean; and soil- and seed-borne diseases in several crops. In cereals (wheat, barley, rye, oat) and vegetables, several diseases can be present either together as complex or separated by regions, varieties or one after the other, whereas only few diseases are important in crops such as maize and oilseed rape (Fig. 8.1a).

8.4 What Is the Spectrum of Activity for Fungicides?

A similar approach as for the Disease – Crop Matrix was taken for the Disease – Fungicide Matrix (Fig. 8.2) illustrating the importance and spectrum of activity for fungicides. The ranking of the different fungicide classes was done by fungicide market size, which in 2006 was 28% for DMIs (demethylation inhibitors, mainly triazoles, imidazoles), 19% for QoIs (quinone outside inhibitors, e.g. azoxystrobin, pyraclostrobin), 17% for multisite inhibitors (such as mancozeb, chlorothalonil), 4% for benzimidazoles (e.g. benomyl, thiophanate-methyl), 3% each for dicarboximides (e.g. iprodione), phenylamides (e.g. mefenoxam), amines (e.g. fenpropidine), anilinopyrimidines (e.g. cyprodinil), phenylpyrroles (e.g. fludioxonil) and MBIs (melanine biosynthesis inhibitors, e.g. pyroquilon, carpropamid), 2% each for carboxamides (SDHIs, succinate dehydrogenase inhibitors such as carboxin, boscalid), CAAs (carboxylic acid amides, e.g. dimethomorph, mandipropamid), and plant defence inducers (induced resistance) such as acibenzolar-S-methyl (Bion) (values in % of total sales; Syngenta internal data). The level of

a

	Diseases	biotrophic			perthotrophic				bacteria
		downy mildew[3)]	powdery mildew	rust	fruit rots & moulds	leaf spots	soil borne	seed borne	
crops		1	2	3	4	5	6	7	8
A	Cereals		++	++		++	+	++	
B	Rice					++	++	+	++
C	Grape	++	++		++				
D	Pome & stome fruits	+	++		++	++			+
E	Soybean	+		++		++			
F	Sugarbeet		+			++	+	+	
G	Potato	++				++	+	+	
H	Vegetables[1)]	++	++	+	++	++	++	+	+
I	Oilseed-rape					+	+		
K	Peanut			+		++	+		
L	Cotton					+	++	+	+
M	Tropical fruits[2)]	++		++	++	++	+		
N	Turf & ornamentals	++	++	+		++	+		+
O	Maize	+				+	++	+	

Fig. 8.1 (**a**) Disease – Crop Matrix, sorted by market size of products for disease control: extent of damage is ++ major loss; + minor loss; blank – not existing or economically not important. (**b**) Disease – Crop Matrix: Index with names of pathogen genera (next page)

b

A2	Erysiphe (Blumeria)
A3	Puccinia spp.
A5	Septoria(Mycosphaerella/ Leptosphaeria) (wheat) Pyrenophora, Rhynchosporium (barley)
A6	Pseudocercosporella, Fusarium, Microdochium, Gaeumannomyces
A7	Ustilago (barley), Tilletia, Fusarium (wheat)
B5	Pyricularia (=Magnaporthe), Cochliobolus
B6	Pellicularia (=Rhizoctonia), Pythium, Microdochium
B7	Gibberella (=Fusarium)
B8	Xanthomonas
C1	Plasmopara
C2	Uncinula
C4	Botrytis
D1	Phytophthora (apple)
D2	Podosphaera (apple)
D4	Botrytis, Monilinia, Gloeosporium
D5	Venturia (pome), Stigmina (stone)
D8	Erwinia, Pseudomonas (stone)
E1	Peronospora, Phytophthora
E3	Phakopsora
E5	Septoria, Diaporthe, Colletotrichum

F2	Erysiphe
F5	Cercospora, Ramularia
F6	Polymyxa/Rizomania-Virus
F7	Phoma
G1	Phytophthora
G5	Alternaria
G6	Rhizoctonia, Helminthosporium
G8	Erwinia, Clavibacter, Ralstonia, Streptomyces
H1	Phytophthora, Peronospora, Pseudoperonospora
H2	Sphaerotheca, Erysiphe
H3	Uromyces (beans, pea)
H4	Botrytis
H5	Alternaria, Colletotrichum, Botrytis
H6	Rhizoctonia, Ascochyta, Fusarium, Verticillium, Pythium
H7	Colletotrichum
H8	Erwinia, Pseudomonas, Xanthomonas
I5	Sclerotinia, Alternaria
I6	Rhizoctonia, Pythium, Leptosphaeria (=Phoma)
K3	Puccinia
K5	Mycosphaerella (=Cercospora)
K6	Sclerotium, Rhizoctonia, Aspergillus

L5	Colletotrichum (=Glomerella), Alternaria, Ascochyta, Ramularia, Cercospora
L6	Rhizoctonia, Verticillium, Fusarium, Pythium
L7	Aspergillus
L8	Xanthomonas
M1	Phytophthora (citrus, cacao)
M3	Hemileia (coffee)
M4	Colletotrichum (=Glomerella) (banana), Penicillium (citrus), Phytophthora (cacao)
M5	Mycosphaerella (banana, coffee), Diaporthe (citrus)
M6	Phytophthora (citrus)
N1	Peronospora
N2	Erysiphe
N3	Puccinia
N5	Helminthosporium, Colletotrichum, Sclerotinia
N6	Rhizoctonia, Fusarium, Pythium, Sclerotium
N8	Pseudomonas, Xanthomonas
O1	Peronosclerospora
O5	Helminthosporium (Drechslera, Cochliobolus)
O6	Fusarium, Pythium
O7	Ustilago, Sphacelotheca

Fig. 8.1 (continued)

	Diseases	biotrophic			perthotrophic				bacteria
		downy mildew	powdery mildew	rust	Fruit rot & moulds	leaf spots	soil borne	seed borne	
Mode of action groups FRAC code 1)		1	2	3	4	5	6	7	8
G1	DMIs		++	++	+	++	+	+	
C3	QoIs	++	++	++	++	++	+	+	
M	Multisites	+	+		+	+	+		+
B1	Benzimidazoles		+		++	++	+	+	
E3	Dicarboximides				++	+	+		
A1	Phenylamides	++					+	+	
G2	Amines		++	++	+				
D1	Anilinopyrimidines				++	++	+		
E2	Phenylpyrroles				++		++	++	
I	MBIs (Melanin-Inh.)					++			
C2	Carboxamides				+	+		+	
F5	CAAs	++							
P	Induced resistance	+	+						+

Fig. 8.2 Disease – Fungicide Matrix, sorted by fungicide market size: Level of disease control is ++ excellent; + moderate; blank – no effect

disease control was defined with two plus (++) for excellent control, one plus (+) for moderate control and blank for no control. The fungicide classes are named according to FRAC nomenclature (www.frac.info; Publications, FRAC Code List, 2008).

Some fungicides have a large spectrum of activity such as DMIs (G1/2, 3, 5) and QoIs (C3/1, 2, 3, 4, 5), others are very specific with a rather narrow spectrum of activity such as phenylamides (A1/1), CAAs (F5/1) and MBIs (I/5) (Fig. 8.2).

8.5 How to Study and Explain Inheritance of Traits in Pathogens

The understanding of genetics and molecular processes in pathogens is of vital importance. Such phenomena can be explained simply by using fungicide resistance as a marker to follow inheritance of traits in fungi (which are mostly haploid) and related organisms like oomycetes (which are diploid). Fungicide resistance can be based on three major mechanisms: (1) mutation(s) in the target gene resulting in amino acid changes in the protein (enzyme pocket); (2) overexpression of the target enzyme; and (3) the fungicide does not reach the target site in effective concentrations either through enzymatic degradation of compounds (metabolic resistance; very rare for fungicides) or through increased efflux of the compound from the cell, for example through up-regulation of transporter genes and ABC pumps. For fungicides, the first possibility, mutation(s) in the target gene, is by far the most important mechanism. The target (resistance) genes and mutations in these genes can be used as genetical or molecular markers for diagnostics (qualitative or quantitative PCR) and to follow the inheritance of traits.

Especially interesting is the segregation pattern of resistance to fungicides such as phenylamides (PAs, e.g. mefenoxam, Fig. 8.3) and carboxylic acid amides (CAAs, e.g. mandipropamid, Fig. 8.4) in diploid plant pathogens such as *Phytophthora infestans* and *Plasmopara viticola*, because Mendelian rules can be verified. When a PA-sensitive and a PA-resistant isolate of *P. infestans* was crossed, all F1 (oosporic) progeny isolates were intermediate in sensitivity to mefenoxam (0.01 mg/L < EC 50 < 10 mg/L) except for one sensitive and two resistant offsprings which were selfs of the respective parents (Fig. 8.3a). In a cross of two F1 isolates which were intermediate in sensitivity, the segregation of sensitive (EC 50 < 0.15 mg/L), intermediate (0.15 mg/L < EC 50 < 10 mg/L) and resistant (EC 50 > 10 mg/L) offsprings in the F2 progeny was s:i:r = 4:38:4 (approximately 1:9:1) (plus two intermediate selfs each of the two parents) (Fig. 8.3b). This segregation pattern suggests the presence of one semi-dominant gene for PA-resistance (expected segregation pattern 1:2:1) influenced by the action of 'minor' genes (Gisi and Cohen 1996).

When a CAA-sensitive and a CAA-resistant isolate of *P. viticola* was crossed, all F1 (oosporic) progeny isolates were sensitive to mandipropamid (EC 50 < 5 mg/L, Fig. 8.4a). This was surprising since no (phenotypic) segregation of resistance was observed. In a cross of two F1 isolates which were (phenotypically) sensitive, the segregation of resistance in the F2 progeny was s:r = 62:7 (approximately 9:1) (EC 50: s < 5 mg/L; r > 100 mg/L) (Fig. 8.4b). This segregation pattern suggests the presence of one (or two) recessive nuclear gene(s) for CAA-resistance (expected segregation pattern 1:3 for one gene, and 1:7 or 1:15 for two genes with one or two heterozygous loci, respectively) (Gisi et al. 2007). It was also demonstrated with the same isolates that resistance to all CAA fungicides (e.g. mandipropamid, dimethomorph, iprovalicarb) co-segregated in F2, which is genetic proof of cross-resistance. However, resistance to CAAs did not co-segregate with resistance to PAs underlining that no cross-resistance is present between CAA and PA fungicides (Gisi et al. 2007).

The segregation pattern of PA- and CAA-resistance described above for *P. infestans* and *P. viticola* are classic examples of how easy it is to demonstrate genetic processes in agronomically important plant pathogens in teaching programmes, a subject which is especially interesting for more advanced students (Master and Ph.D. candidates). In addition to the examples given above for resistance genes (resistant phenotypes), also specific molecular markers (e.g. neutral markers like simple sequence repeats, SSR, or single nucleotide polymorphisms, SNPs, like the G143A mutation in the cytochrome b gene) can be used for studying and explaining inheritance and segregation processes in plant pathogens (Knapova et al. 2002; Gisi and Sierotzki 2008).

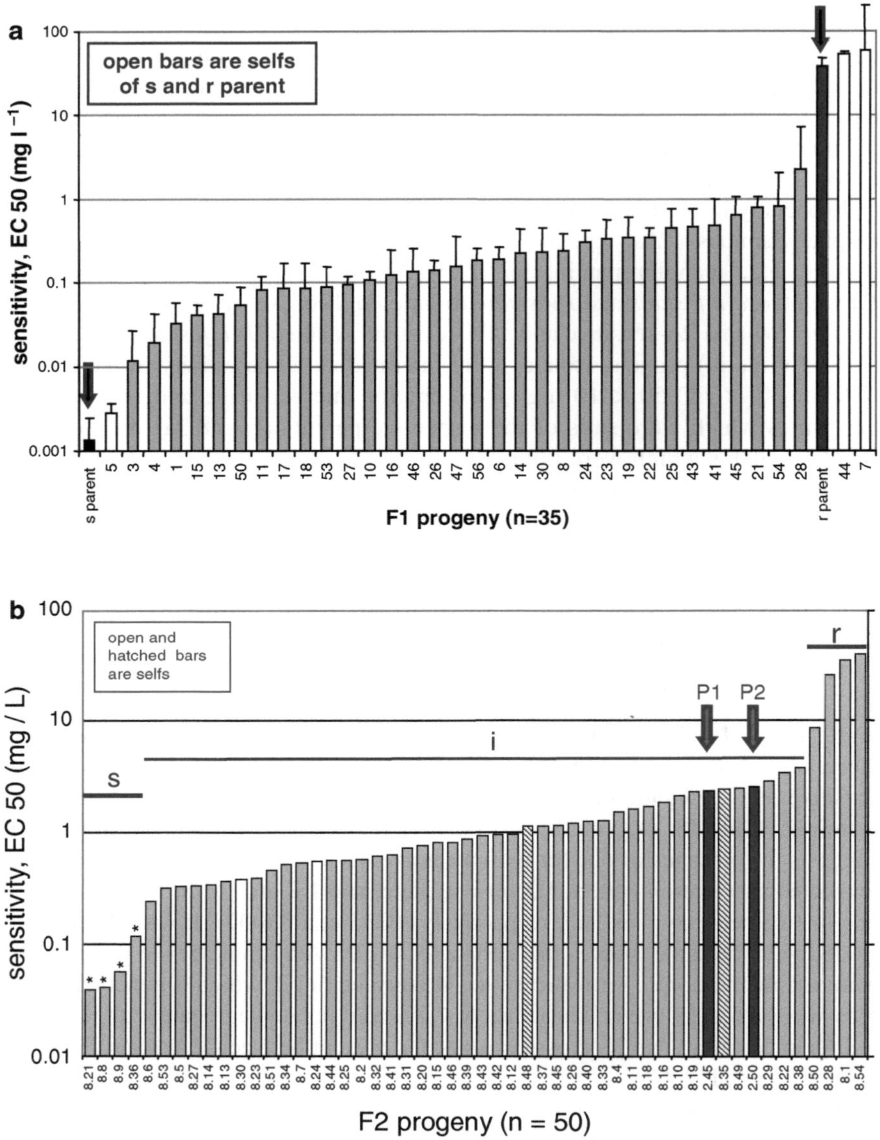

Fig. 8.3 (**a**) Segregation of sensitivity (EC 50 ± SD) to mefenoxam in F1 progeny generated from a cross with sensitive and resistant parents (F0, *black bars* marked by *arrows*) in *Phytophthora infestans* (Knapova et al. 2002). (**b**) Segregation of sensitivity (EC 50) to mefenoxam in F2 progeny generated from a cross with intermediately sensitive parents P1 and P2 (F1, *black bars* marked by *arrows*) in *Phytophthora infestans* (Knapova et al. 2002)

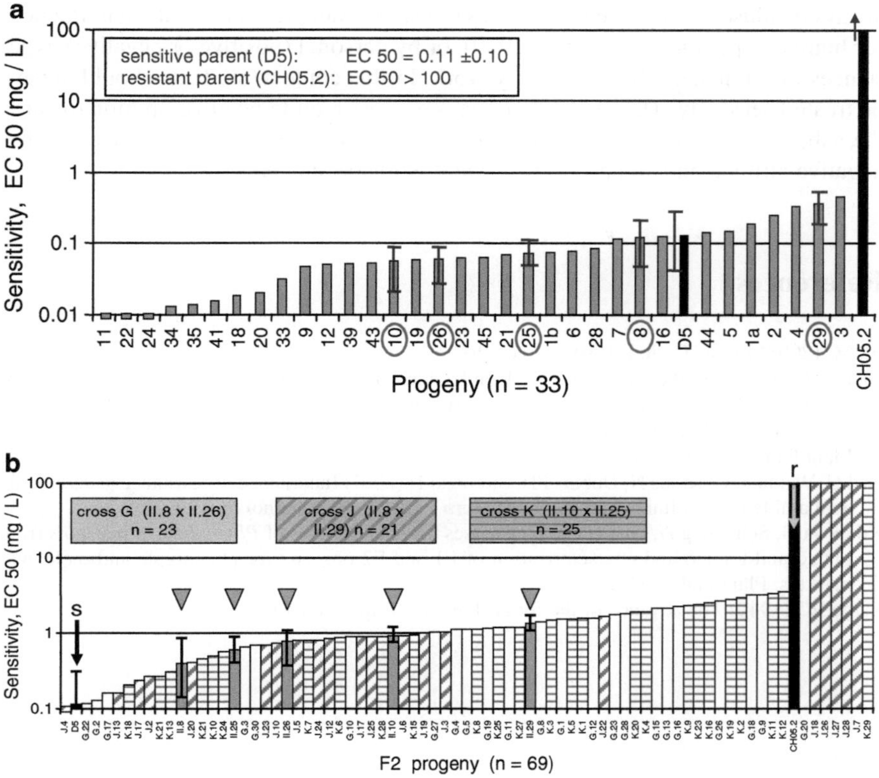

Fig. 8.4 (**a**) Segregation of sensitivity (EC 50 ± SD) to mandipropamid in F1 progeny generated from a cross with sensitive and resistant parents (F0, *black bars*) in *Plasmopara viticola*. Offsprings marked with *circle* were used as parents for generating F2 progeny (Gisi et al. 2007). (**b**) Segregation of sensitivity (EC 50) to mandipropamid in F2 progeny generated from three crosses (G, J, K, *white* and *hatched columns*) with sensitive F1 parents (*grey columns* marked with *triangles*) in *Plasmopara viticola* (*black columns* marked with *arrows* are F0 parents) (Gisi et al. 2007)

8.6 Conclusions

In this paper, four aspects of plant pathology were identified as being especially suitable for teaching purposes: (1) most important symptoms of plant diseases; (2) most important diseases on agronomically important crops (using a disease – crop matrix including nomenclature of pathogen genera); (3) spectrum of activity for fungicides used against most important diseases (using a disease – fungicide matrix including nomenclature of fungicide modes of action); and (4) inheritance of genes and molecular markers, by using fungicide resistance as an example. Because of its complexity, teaching plant pathology remains a major challenge for both teachers and students. Whether the teaching programme is oriented to more fundamental (academic) subjects or to agronomically relevant aspects, training courses should always be illustrated by selected case histories, for example events with consequences for human population (e.g. Large 1940), or by personal (positive or negative) experiences of the teacher from his daily work. In this context, students should always be treated seriously. They may sometimes express a quite positive opinion towards the subject and accept what the teacher tries to explain but may also have a rather negative attitude and so have to be treated with special care.

References

FRAC (2008) FRAC Code list. www.frac.info

Gisi U, Cohen Y (1996) Resistance to phenylamide fungicides: a case study with *Phytophthora infestans* involving mating type and race structure. Annu Rev Phytopathol 34:549–572

Gisi U, Sierotzki H (2008) Fungicide modes of action and resistance in downy mildews. Eur J Plant Pathol 122:157–167

Gisi U, Waldner M, Kraus N, Dubuis PH, Sierotzki H (2007) Inheritance of resistance to carboxylic acid amide (CAA) fungicides in *Plasmopara viticola*. Plant Pathol 56:199–208

Knapova G, Schlenzig A, Gisi U (2002) Crosses between isolates of *Phytophthora infestans* from potato and tomato and characterization of F1 and F2 progeny for phenotypic and molecular markers. Plant Pathol 51:698–709

Large EC (1940) The advance of the fungi. Jonathan Cape, London

Chapter 9
Sustainable Crop Protection and Environment Protection: Eight Years of Technology Transfer Between China and Italy

M.L. Gullino, M. Pugliese, N. Capodagli, and A. Garibaldi

9.1 Introduction

At present, agriculture faces the need to meet new challenges both in highly industrialised countries as well as in developing ones. Such challenges are represented by sustainable growth, social integration of rural communities and proper use of the advantages deriving from emerging global markets. A shift toward sustainable agricultural systems, which are more complex in terms of biodiversity, is very important in emerging countries such as China, where 60% of the population (corresponding to 1.3 billion people) lives in rural areas, in poor conditions and still rely on agriculture as the main source of income (OECD 2005). In the effort to reconcile economic and social needs and environmental protection, China is undertaking countermeasures towards the promotion of sustainable agricultural practices. The attention paid to activities in the agro-environmental sector has been increasing over time due to the high social and economic priority attached by the Chinese authorities to the modernization of agriculture that must be pursued in a sustainable manner, addressing at once food security, environment protection, economic development, and good management of natural resources. On one side, China is committed to comply with the multilateral environmental conventions and protocols with direct impact on the agricultural sectors (e.g. the Montreal Protocol on Substances that Deplete the Ozone Layer, the Stockholm Conventions on Persistent Organic Pollutants, the Convention on Biological Diversity, the Convention to Combat Desertification, the Framework Convention on Climate Change); on the other side it is pushing the adoption of agricultural practices and technologies with low impact on the environment. Crop protection is an important component of sustainable

M.L. Gullino (✉), M. Pugliese, and A. Garibaldi
Centre of Competence for the Innovation in the agro-environmental sector (AGROINNOVA), University of Torino, Via Leonardo da Vinci 44, 10095, Grugliasco (Torino), Italy
e-mail: marialodovica.gullino@unito.it

N. Capodagli
Huayimeng Environment, Science & Technology Consulting Co., Ltd, Beijing

N.V. Hardwick and M.L. Gullino (eds.), *Knowledge and Technology Transfer for Plant Pathology*, Plant Pathology in the 21st Century 4,
DOI 10.1007/978-1-4020-8934-3_9, © Springer Science+Business Media B.V. 2010

agriculture. The results of a 8 years co-operation between Italy and China in the field of sustainable crop protection are discussed.

9.2 A Brief Overview of Agriculture in China

China is a large agricultural country. Given its size and importance, the agricultural economy of China commands considerable attention, playing an expanding role in world trade. Although the agricultural sector is declining, it still represents an important element of China's economy. Agriculture accounts for almost 15% of the GDP and above 40% of employment. Agricultural production structures are dominated by small-scale farming: about 200 million farms with an average land area of just 0.65 ha per household (OECD-FAO 2008).

The transformation of China's agricultural sector is crucial to achieve the country's goal of building a stable and productive society. China's performance in the agricultural sector would not only affect the world's food and agricultural markets but also have a profound impact on the global war against poverty (Dong et al. 2006).

China's rapid economic expansion appears to have finally exhausted the pool of under-employed workers. Since 2003, wages have been rising at a double-digit pace. The dwindling pool of available rural workers is resulting in increased mechanisation of harvesting and planting. Anecdotal evidence also suggests that intensive agricultural practices, such as double-cropping, transplanting seedlings by hand, and small-scale pig production, have decreased due to labour shortages and high wages (Lohmar and Gale 2008). Now, according to the Ministry of Commerce of China (MOFCOM), the peasant income maintained the fourth consecutive year of rapid increase. After deducting the price index and being compared with that of 2006, the net farmer income increased by 9.5%. This has been the highest growth rate since 1997 ('Invest in China' 2007 Survey).

The migration of young farm workers to cities may halt the ongoing conversion to high value crops (i.e. vegetables, fruit, and grapes) usually more labour-demanding. During the last 15 years the area of fruit and vegetable production expanded by an average of 1.3 Mha per year. In a country where 60% of the population lives in rural areas, in poor conditions, and still relies on agriculture as main source of income, the shifting to high-value crops is another option for achieving a higher quality of life (Brown 2005).

9.2.1 Farm Inputs

China's current exploitation of land and water resources is either at, or beyond, sustainable levels. The cultivation of steep hillsides is causing massive sedimentation loss estimated at over 2 Gt per year, decreasing productivity in areas losing topsoil, reducing water storage capacity in reservoirs, and increasing the likelihood of floods. Agricultural practices, both crop cultivation and animal husbandry, on sensitive arid grasslands, are partly to blame for the desertification of these areas. In the North China Plain, the groundwater table is falling rapidly in some areas, and several surface-water sources periodically dry up before reaching the sea (Lohmar and Gale 2008).

China is the world's biggest user, producer and exporter of pesticides. According to Xinhua, China's main state news agency, the annual pesticide use in China is about 1.2 Mt, on approximately 300 Mha. According to MOA, the volume of farm pesticide application was over 1.4 Mt in 2006 (MOA 2007 report). Official statistics show that China produces about 300 types of pesticides and an additional 800 types of pesticide mixtures. In 2006, China produced 1.3 Mt of pesticides and exported 583,000 t.

Fertiliser utilisation is also being improving in recent years in China. In line with the scientific outlook on development, efforts were made to promote the development of resource-saving and environment-friendly agriculture and shift the mode of growth in crop farming from extensive to intensive. According to statistics of MOA, the volume of chemical fertiliser application (net volume) was over 47 Mt in 2006. Balanced fertilisation, which was warmly welcomed by farmers, made much headway in 2006 with the vigorous support of financial and other government departments and the concerted efforts of agricultural sectors at all levels, effectively raising the rate of fertiliser utilisation. Soil testing and formula fertilisers were applied on a total of 17.3 Mha of farmland in the nation's 600 counties in 2006, reducing the use of about 500,000 t of fertilisers, raising the chemical fertiliser utilisation rate by 5% and reducing the cost of production each hectare by 375 yuan on average. In 2006, China produced 55.93 Mt of fertilisers, exporting 5.41 Mt, while importing 11.29 Mt (MOA 2007).

Chinese farmers have applied heavy doses of chemical fertiliser and pesticides to overcome natural resource constraints and significant pest pressures. Farmers use a variety of veterinary drugs to control diseases that spread quickly among livestock and fish raised in crowded facilities, and they use feed additives to enhance animal growth. Residues of toxic pesticides, drugs, and industrial pollutants detected in food are a potential health hazard. A sizeable share of China's industrial production also takes place in rural areas and in close proximity to agriculture. The external costs of industrial production, such as water pollution, are often borne by agricultural producers (Lohmar and Gale 2008).

9.2.2 Productivity

According to statistics of MOA, in 1990 grain accounted for 79.3% of area sown in China, while vegetables only 4.4%. In 2006 grain area was 68.2% and vegetables 11.8%. Major cash crop production reached a historical high in 2006. Cotton output reached 6 kg more than the record figure of 2002. Rapeseed, groundnuts, sugar beet/cane outputs all increased prudently. Steady increases were also reported in the production of vegetables, fruit, tea, flowers and edible fungi. Total output of fruit in 2006 was over 165 Mt, while it was below 25 Mt in the beginning of the Nineties. In 2006, total livestock output value reached 1,364 billion yuan, accounting for 32.2% of the total agricultural output value, with 15.3% of increase of dairy products (MOA 2007). In 2006, according to MOA, the composition of value added of the agriculture sector was as follows: crops 56.4%, forestry 4.4%, livestock 26.6%, and fish 10.3%, while in 2003 it was 55.6%, 4.8%, 26.8% and 10.3%, respectively.

While China has emerged as the world's leading importer of soybeans, vegetable oil, cotton, wool, rubber, and animal hides, it has been surprisingly successful at meeting the basic food needs of its population of more than 1.3 billion people and it has stepped up as a major food exporter. How long can China sustain this momentum?

China imports only small amounts of premium-grade rice, minor amounts of wheat in most years, and (almost) no maize. China has maintained agricultural self-sufficiency in grains as it carries out the world's largest and fastest urbanisation and industrialisation. Economic development is increasing competition for scarce resources in China, but growing incomes are allowing most consumers to increase consumption of fruit, vegetables, and livestock products. China dominates world markets in a variety of products areas, including garlic, apples, apple juice, mandarin oranges, farm-raised fish and shrimps, and vegetables.

9.3 Agricultural Research in China

China's challenge to develop sustainable solutions to social and economic pressure put on agriculture is even greater due to the threat posed by agriculture on the environment and human health. China's agriculture represents one of the most polluting production sectors. After China's accession to the WTO, safety concerns on Chinese products have been raised as potential barrier to the prospected increasing export of agro-food to Western countries, Europe in particular, because of its high quality and sanitary standards (Rozelle and Huang 2006).

China's food industries have been stung by quality and safety problems both overseas and in domestic markets. There is a strong campaign to reduce and regulate farm chemical use. Chinese officials now ban food production in heavily polluted areas and limit use of toxic chemicals. Exporters must go through stringent certifications and product testing, raising the costs of production and limiting the development of potential export markets for food products (Lohmar and Gale 2008).

9.3.1 The Tenth Five-Year Plan (2001–2005)

The Tenth Five-Year Plan (2001–2005) allocated more than US$950 million to the improvement of agricultural infrastructures, the promotion of basic and applied research, the conversion to innovative technologies, the training of extension workers and growers.

While priority were given to boosting grain production and to stop the increasing reliance on grain imports, particular attention was paid to keep the positive trends registered during the past 10 years in fruit and vegetable production. Indeed, the area invested in fruit and vegetables passed from 10 Mha in 1991 to 26 Mha in 2003, mainly in response to a rapid growth in domestic demand and in the export market. In a country like China, where the average farm area is around 0.65 ha, the shift to higher value crops is a good solution for increasing wages. Currently the total area cropped with fruit and vegetables is about 28 Mha, reached during year 2006 (www.agri.gov.cn). Because of the policies on self-sufficiency included in the 11th Five-Year Plan of China, according to experts of the Ministry of Agriculture, this is considered to be the peak of the area expansion in these sectors.

Building on one of the strongest research systems in the world, during the 1980s and 1990s, China's agricultural scientists and the extension system developed and disseminated technology throughout the People's Republic Period. Breeders during the Reform era turned out a constant stream of rice, wheat, and maize cultivars. Also, during the same period, breeding effort enhanced the quality of its seed stock. The government also made a strong commitment in plant biotechnology and, for instance, China's research community has made a major investment into understanding the structure and function of the rice genome and methods of transforming several crops. Although China only started its national plant biotechnology programme in the mid-1980s, a number of years after most other developed countries, it has grown quickly and taken a relatively unique path in the world, with the majority of plant biotechnology research being funded by the government (Rozelle and Huang 2006).

9.3.2 The 11th Five-Year Plan (2006–2010)

The 11th Five-Year Plan (2006–2010) is giving greater consideration to crop protection and reduction of use of pesticides, as well as food safety and the scientific and technological content of the agriculture development. Continuing the policy of modernisation of the agriculture sector started during previous Five-Year planning periods, China will continue to consolidate, improve and strengthen its policies in support of agriculture and increase by a large margin (not less than 30–40%) its investment in agriculture, the rural economy and rural areas (MOA 2007). The government priority is given to national food security and grain self-sufficiency, which has an overall bearing on economic and social development and affects the vital interest of the people, so the area sown to grain crops will be effectively kept stable and the yield per unit area increased, and the government will increase direct subsidies for grain producers. In order to guarantee the national food security (which is intended as 'self-sufficiency') and for continuing to have 'an adequate stock of grain', as well as sufficient other basics, such as vegetable oil and meat as well as other commodities in short supply, the government will also implement policies and measures as a matter or urgency to support production and ensure co-ordination in the production, transport and sale of products. During the Five-Year implementation period, China will adhere to the strictest possible system for protecting farmland, and in particular increasing protection of basic farmland. The government will carefully examine and adjust the amounts and standards for land use in all types of plans in accordance with the master plan for land use, and strictly control it, as well as resolutely put a stop to illegal appropriation of arable land and forested areas (Wen 2008).

The 11th Five-Year Plan is paying much attention to sustainable crop protection, its success in improving agriculture production, safety and quality being directly dependent on the level of application of modern science and technology. In this regard, Premier Wen Jiabao stated that China will improve the system for spreading agricultural science and technology and providing agricultural technical services, working harder to make innovations in this sector and apply advances in agricultural science and technology, improving the diverse array of mainly non-profit-making agricultural technical services. The government will accelerate agricultural mechanisation, improve the systems for breeding better plant cultivars, information transfer, and the quality and safety of agricultural products. It will also aim to prevent and mitigate disasters, control animal epidemics, plant diseases and insect pests. Trials on soil testing to determine appropriate fertiliser treatments will be increased (Wen 2008), in order to optimise production, save energy and protect the environment, as well as promoting the independent innovativeness and the scientific and technological content (MOA 2006).

9.3.3 Plant Protection

Plant protection has become an important profession and subject for study in China. Plant protection, quarantine and research institutions have been established in agricultural departments and in most of the academies of agricultural sciences above the county level, most of the agricultural schools and universities provide this special field of study and most of the townships have technicians qualified in plant protection. There are 46,000 technicians in plant protection and more than 5,000 researchers specialised in the subject throughout the country. Regulations on plant quarantine and the use of pesticides have been promulgated and relevant standards established, along with emergency plans for dealing with several dangerous plant pests. Funds totalling 2.58 billion yuan were allocated by the state from 1998 to 2006 to build basic facilities for quarantine and prevention, monitoring and early warning, and emergency control of plant pests. At the same time, the central government increased its financial support year by year for major plant disease and insect pest prevention and control. In 2006, 272 million yuan were allocated from the central budget to monitor plagues of locusts (*Locusta migratoria manilensis*), yellow rust (*Puccinia striiformis*) and the larvae of the snout moth (*Crambus agitatellus*). All these efforts greatly improved the means of plant protection and working conditions and strengthened overall protective capacity.

Agricultural departments at all levels continued to regulate pesticides registration and supervise distribution. They tried to prevent the production and use of counterfeit and shoddy pesticides and vigorously promoted the work to reduce and substitute the use of highly toxic pesticides. In 2006, the proportion of highly toxic and deadly pesticides was reduced to 11.8% from 21.8% and environment-friendly pesticides rose from 3.1% to 18.1%. Improvements in the safe use of pesticides are continuing through a mixed approach of legislation and technology transfer.

9.4 Technology Transfer Between China and Italy in the Field of Sustainable Crop Protection

9.4.1 Background

A number of projects in the field of sustainable crop protection have been implemented within the Sino-Italian Co-operation Programme for Environmental Protection, a framework programme jointly launched in 2000 by the Italian Ministry for Environment, Land and Sea and China State Environmental Protection Administration (Gullino et al. 2006; Clini et al. 2008). All projects implemented (Table 9.1) were in response to the primary goal of reducing China's reliance on a massive use of pesticides that is posing serious threats to environment as well as to food safety.

Since the launch of the Sino-Italian Co-operation Programme, significant investments have been made for the phasing out of methyl bromide, a highly toxic fumigant used in the horticultural sector for pre-plant soil sterilisation and banned by the Montreal Protocol because of its implication in the ozone layer depletion (Gullino et al. 2003). In the 1990s, Italy, ranked first in Europe in horticultural crop production and second in the world in methyl bromide consumption, gained considerable experience in developing suitable and feasible alternatives to methyl bromide and invested in transferring the technologies developed to other countries, such as China (Gullino et al. 2003). Preliminary trials, aimed at demonstrating the technical and economic feasibility of innovative and low environmental impact techniques for soil sterilisation, started in 2001 in Shandong and Hebei provinces. The object was to identify solutions which would be applicable in other areas of China. The results eventually contributed to China's Methyl Bromide National Phase-out Plan under the framework of the Multilateral Fund of Montreal Protocol. Hebei and Shandong provinces were targeted because of their high methyl bromide consumption and because they were characterised by an expanding horticultural sector. The selection of target technologies also took into account the local level of infrastructure, mechanisation, availability of agricultural inputs and knowledge. Solutions like soil steam pasteurisation and soil-less cultivation systems were ruled out in favour of cheaper alternatives, which were easier to apply and less energy consuming. Soil solarisation, the use of grafting onto resistant rootstocks and the application of

Table 9.1 Projects on sustainable crop protection implemented within the Sino-Italian Cooperation Programme for Environmental Protection in the period 2000–2008

Project title (duration)	Funded by	Institutions companies involved[a]
Alternatives To The Use of Methyl Bromide in Soil Fumigation (2001–2003)	IMELS	SEPA/FECO; CAU; CAAS; AGROINNOVA
Strengthening Technology and Capacity of Sustainable Agriculture in China (2002–2005)	IMELS	SEPA/FECO; Chinese Research Academy of Environmental Sciences; AGROINNOVA
Sustainable plant protection in respect of the environment: modern techniques for the control of plant pests and diseases of horticultural crops in China (2005–2007)	MAP, ICE, CRUI	AGROINNOVA; CAU, Intrachem Bio Italia SpA, Nuovo Centro S.E.I.A. SpA
Organic Farming Systems and Techniques for the Promotion of "Green" Agriculture in Dongtan Chongming Island (2005–2008)	IMELS	AGROINNOVA; Shanghai Environmental Protection Bureau; Shanghai Academy of Environmental Sciences; SIIC Dongtan Investment & Development (Holdings) Co., ltd.
Innovative techniques for reduction and recycling of agricultural wastes (2006–2009)	MAE, MOST	AGROINNOVA; CAU
Technological innovations in crop protection to enhance food quality in China (2006–2008)	MIUR, IMELS, University of Torino	AGROINNOVA; CAAS; CAU
Anaerobic digestion and composting of agricultural, urban and industrial wastes for the valorization of energetical and agronomical use of biomasses in an ecological italian style "Ecofarm" on Chongming Island (2008)	ICE, MAP	AGROINNOVA; Shanghai Academy of Environmental Sciences; MARCOPOLO Engineering SpA

[a]IMELS Italian Ministry for Environment, Land and Sea, SEPA/FECO Foreign Economic Cooperation Office of the State Environmental Protection Administration, CAU China Agricultural University, CAAS Chinese Academy of Agricultural Sciences, AGROINNOVA Centre of Competence for the Innovation in the Agro-Environmental Sector of the University of Torino, MAP Italian Ministry for Production Activities, ICE Italian Trade Commission, CRUI Conference of Rectors of Italian Universities, MAE Italian Ministry of Foreign Affairs, MOST Chinese Ministry of Science and Technology, MIUR Italian Ministry of Education, University and Research

less harmful chemicals at reduced dosages via drip irrigation, tested on tomatoes and strawberries, resulted of higher acceptance by local growers because, while providing levels of effectiveness comparable to methyl bromide, they required lower cost of investment and smaller changes to fit within the traditional cultural practices (Cao et al. 2002a, b; Dong et al. 2007). For these reasons they were registered by the Chinese Ministry of Science and Technology as successful cases suitable to Chinese agriculture for the control of soil-borne pathogens and as effective alternatives to the use of methyl bromide.

9.4.2 Recorded Examples

Field experiments were conducted in Shouguang County, Shandong Province, east of China, in 2006 and 2007 under the Sino-Italian Co-operation Programme.

9.4.2.1 Control of Root-Knot Nematode

In one experiment, three tomato rootstocks (Beaufort F1, Energy F1 and He-Man F1) were assessed for control of root-knot nematode (*Meloidogyne incognita*) under greenhouse conditions. According to Cao et al. (2008), the results showed that the three rootstocks reduced about 90% the root-knot incidence compared to a susceptible control (FA189) and yields were enhanced by 16–20% (Table 9.2).

In another experiment, three alternatives to MB were compared in greenhouse: grafted rootstock (He-man, *L. lycopersicum x L. hirsutum*), calcium cyanamide and neem oil. Grafting was shown to control root-knot nematodes and increase tomato yields by up to 61.5% compared with the control. The results were comparable to those obtained with MB (Table 9.3). The use of grafting was economically feasible and the net profit ranked as grafting (US$16,480/ha) > MB (US$15,924/ha) > calcium cyanamide (US$14,412/ha) > neem oil (US$8,624/ha) (Wang et al. 2008).

Grafting onto resistant rootstocks is considered a realistic option by Chinese growers, not only for tomatoes but also for cucurbits. According to Davis et al. (2008), China produces more than half of the world's watermelons and cucumbers, and approximately 20% of these are grafted.

Table 9.2 Growth characteristics, root-knot index and yields of tomato plants grown on three resistant rootstocks compared with the variety FA189 (Shandong Province, P.R. China, 2007) (Adapted from Cao et al. 2008)

Grafted plant	Height (cm)	Dry mass of stem (kg * plant^{-1})	Dry mass of root (g * plant^{-1})	Root-knot index	Total yield (kg * m^{-2})
Beaufort	133.03 b[a]	0.11 a	12 b	13.75 a	16.22 a
Energy	140.83 a	0.11 a	12 b	13.13 a	16.08 a
He-man	136.99 ab	0.10 a	11 b	5.94 a	15.78 a
Control (FA189)	142.80 a	0.08 b	36 a	100 b	13.58 b

[a]Different letters in the same column mean significance at $P < 0.05$ level by Duncan's test

Table 9.3 Evaluation of three alternatives to methyl bromide in terms of root-knot incidence, yields of tomato plants and economical profit (Shandong Province, P.R. China, 2007) (Adapted from Wang et al. 2008)

Treatment	Control (FA-189)	Methyl bromide	Calcium cyanamide	Neem oil	He-man
Root-knot index	100%	0	64%	96%	4%
Yield of tomato fruit (10^4 kg ha^{-1})	10.75 c[a]	17.35 a	16.51 a	13.82 b	17.03 a
Increase production	–	61.5 %	53.6 %	28.6 %	58.5 %
Profit ($ha^{-1})	8,446	24,370	22,858	17,070	24,926

[a]See Table 9.2

9.4.2.2 Drip Irrigation

A similar approach, avoiding the pursuit of short-term objectives due to the pressure for profit, was adopted in different regions, in order to ensure the reproducibility and long-term sustainability of transferred technologies. In Xinjiang and Inner Mongolia, for instance, Chinese western regions characterised by a fragile ecosystem and by poor social conditions, a very low level of infrastructure and scarce capacity in farmers for managing modern cropping systems, it was important to use very basic and low cost technologies. Drip irrigation systems resulted beneficial solutions to the serious problem of desertification, soil erosion and pollution affecting the two regions. Used in substitution for the locally adopted flood irrigation and also for the distribution of fertilisers at reduced dosages, drip irrigation systems achieved significant reduction in the use of water and fertilisers (five to six times less compared to common practices) on tomatoes, pumpkins, cabbages, grapes and maize. These were promising results for regions like Xinjiang, formerly one of the poorest regions of China and now preparing to be one of the main agricultural production areas of the country.

9.4.2.3 Biodegradable Plastic Film

Different considerations should be made on the use of starch-based biodegradable plastic films used in replacement of traditional polyethylene mulching films, cause of 'white pollution' in the area. China usage of plastics in agriculture is the highest in the world (Table 9.4), and this usage increased almost 6,000 times between 1980 and 2000, from 1,600 ha to nearly 10 Mha (Scott 1999), which also dramatically lead to an increase in the pollution of Chinese agricultural soils. The agronomic and environmental performance of biodegradable films was satisfactory: providing good control of weeds and a complete degradation of the film a few months after the end of the cropping cycle were observed (Table 9.5).

However, since the cost of biodegradable plastic is still too high in comparison with traditional plastic films (Table 9.6) their actual transfer into practice will be delayed until the price differential is significantly reduced (Gullino et al. 2002).

9.4.2.4 Cropping Systems

More specialised technologies and complex cropping systems were considered for implementation in Chongming (Shanghai) compared with Xinjiang and Inner Mongolia. The Shanghai area is the most advanced in China for its technology,

Table 9.4 Usage of plastics in agriculture (% of total consumption)

Country	Usage (%)
China	20
Israel	12
Spain	8
USA	4

Table 9.5 Degradation of mulching film and effect on weeds of biodegradable (Mater-Bi) and PE films on watermelon (China, 2007)

Mulching film – thickness (μm)	Degradation index of film at the end of the crop[a] film upon soil[b]		buried film[c]		weeds at 27/07/2007 number/m^2	
Mater-Bi – 15	5.2	b[d]	4.0	b	17.6	a
PE black – 12	7.1	a	9.0	a	16.4	a
Bare soil	–		–		40.6	b

[a]Transplanting: 17/05/2007; end of the crop: 28/07/2007
[b]Degradation index of the film upon the soil (1 = 0% of mulched soil; 9 = 100% of mulched soil)
[c]Degradation index of the buried film
[d]See Table 9.2

Table 9.6 Comparison between the costs of biodegradable films and conventional PE in China

Characteristic of the film	PE	Biodegradable films	
Thickness (μm)	15	15	12
average weight (kg/ha)	150	180	140
Cost of the product (€/ha)	210	900	700
Cost difference (€/ha) (base: PE)	–	690	490
Cost difference (%) (base: PE)	–	76.7	70.0
Average removal cost (€/ha)	40	0	0
Average disposal cost (€/ha)	6	0	0
Overall cost of the product (€/ha)	256	900	700
Overall cost difference (%) (base: PE)	–	71.6	63.4

knowledge, foreign trade and capital turnover. The expectation in terms of technology transfer is quite high.

In Chongming Island, the third biggest Island of China after Taiwan and Hainan a few kilometres from Shanghai, IMELS and the Shanghai Municipal Government implemented a project meant to convert the traditional local agricultural systems into organic farming production, with special focus on crop protection. Chongming Island is the world's largest alluvial island. Its coastal wetland and tidal flats provide many important ecological features, including buffers against tidal surges and staging areas for migratory birds. Due to its extraordinary resources, scenic qualities, and its proximity to the city of Shanghai, 45 km away, the island is also an attractive tourist destination, and it supports important agricultural and fisheries economies. The aim was to develop environmentally friendly green food production not only to increase potential for higher income for local growers looking with interest to foreign markets, but also to enable the production of healthy food and the promotion of a safe environment for national eco-tourists visiting in future Chongming Island. In particular the project aimed at stopping soil salinisation processes, to reduce the use of chemical fertilisers and pesticides.

Field experimental were conducted over 2 years on tomatoes, watermelons, maize, pear and other horticultural crops. The technical and economical feasibility of the use of grafting on resistant rootstocks, biodegradable mulching films in combination with the use of 'fertigation' and environmental monitoring systems, diagnostic kits for plant pathogens and integrated pest management, also based on the use of biological products, were evaluated, as well as their environmental impact and socio-economical benefits.

The new technologies were able to increase product quality and safety, resulting in an increase in incomes and positive economic benefits, while maintaining production costs, in many cases, similar to those of the conventional system (Table 9.7).

9.4.2.5 Pesticides

The use of chemical pesticides, in particular of older generations, no more permitted in Europe, like flusilazol, carbendazim, metalaxyl, fenvalerate and triadimefon, have been replaced by more environmental friendly and biological product. For instance, in the case of pear, the reduction in pesticides usage reached, 29% and 69%, respectively, compared to conventional use, while maintaining similar yields (Table 9.8). In the case of waxy maize, pesticides reduction reached 67% in 2006 and 100% in 2007, while yield increases of 6% in 2006 and up to 15% in 2007 were obtained (Table 9.8). No chemical pesticides were used at all for watermelon and tomatoes in 2007.

Table 9.7 Production costs of new technologies for different horticultural crops on Chongming Island (Shanghai) in 2006 and 2007

	Production costs (RMB/ha)		Difference	
Crop (year)	Conventional technologies	New technologies	(RMB/ha)	Ratio (%)
Pear (2006)	48,150	45,374	−2,776	−6
Pear (2007)	36,814	31,260	−5,554	−15
Corn (2006)	12,416	18,507	+6,091	+49
Corn (2007)	9,818	16,168	+6,350	+65
Watermelon (2007)	20,761	21,693	+932	+4
Tomato (2007)	39,730	38,989	−741	−2

Table 9.8 Comparison of chemical pesticides usage and yields of different horticultural crops on Chongming Island (Shanghai) in 2006 and 2007

	Chemical pesticides (g/ha of active ingredients)			Yield (kg/ha)	
Crop (year)	Conventional technologies	Transferred technologies	Reduction (%)	Conventional technologies	Transferred technologies
Pear (2006)	1,508	1,069	29	15,532 a[a]	15,632 a
Pear (2007)	1,330	415	69	13,540 a	13,024 a
Corn (2006)	791	263	67	9,420 b	10,037 a
Corn (2007)	148.5	0	100	9,663 b	11,437 a
Watermelon (2007)	2,022	0	100	12,576 b	24,582 a
Tomato (2007)	4,815	0	100	22,320 a	23,741 a

[a]See Table 9.2

Table 9.9 Effect of rootstocks PS1313 and FR Strong (Seminis) on yields of two local watermelon varieties in Chongming Island (Shanghai) in 2007

Scion	Rootstock	Yield (kg/ha)
Red pulp	–	12,576 c[a]
Red pulp	PS1313	23,068 b
Red pulp	FR strong	24,582 b
8424	–	19,088 bc
8424	PS1313	25,361 b
8424	FR strong	34,813 a

[a]See Table 9.2

9.4.2.6 Grafting

Grafting on resistant rootstocks was confirmed as a useful technique, being able to control pathogens, reduce problems of soil salinity and also increase yield. Local cultivars grafted on resistant rootstocks were compatible and, in the case of watermelon, yields were increased by 55% compared to non-grafted plants (Table 9.9).

This project represented a fruitful example of co-operation between the private and public sectors in China and in Italy and a model of research and semi-commercial scale application of innovative techniques and technologies for sustainable crop protection. The project went beyond the merely environmental concerns and strengthened the role of rural areas as multifunctional dynamic systems. This is an important aspect in China, since the present economic growth, urbanisation and increased leisure time, also increase the demand for tourism and recreation activities in rural areas.

The use of biocontrol agents and grafting onto resistant rootstocks within integrated pest management systems has been addressed by the project outlined here. Within the Scientific and Technological Co-operation Programme between the Italian Ministry of Foreign Affairs and the Chinese Ministry of Science and Technology, a model of research and semi-commercial scale application of innovative techniques and technologies for organic agricultural waste composting as well as of the use of biodegradable plastics as a means to reduce production of non-compostable agricultural waste is being developed under the project "Innovative techniques for reduction and recycling of agricultural wastes".

9.5 Projects Features and Constraints

While the main aim of the projects was to enable China's compliance with the obligations set by the Multilateral Environmental Agreements and with the Millennium Development Goals, objectives and instruments of implementation tried to be site specific and to address properly the particular social and economic needs of the areas investigated. In large developing countries, like China, with an extremely diversified agricultural sector in terms of climate, levels of infrastructure and mechanisation, economic and social conditions, it was important to avoid generalised approaches.

9.5.1 Crop and Site Selection

The lack of technical and scientific knowledge and capacity in managing innovative cropping systems emerged as one of the major barriers towards the actual adoption of new techniques. The results from the Sino-Italian agro-environmental projects were not surprising, as they reflected the general Chinese situation. Although there is an urgent need to create a higher profile for researchers and extension staff supporting, in the long term, the conversion to sustainable agricultural practices investments in agricultural research and education are low. The Chinese extension system needs to seek a new balance after conversion from the central planning systems to the 'Household Responsibility Systems', in which the individual farmers become the basic production unit, rather than the production task forces of the previous collective systems. The challenge is to re-orientate scientific and technical capabilities towards the market and industry requirements. Again, international co-operation programmes play a fundamental role in filling the educational gap.

The selection of fruit and vegetables as target crops provided a consistent reference point in all the projects, due to environmental and economic reasons. While grains remain the key crop in China, their share of total crop production and in the area sown, declined quite substantially between 1990 and 2003 as other crops, like fruit and vegetables, became more profitable and the government relaxed most of the policy measures, which had previously forced farmers to produce cereals (OECD 2005). However, if changes in the domestic demand and emerging export opportunities lead the impressive increases in vegetable and fruit production and provided farmers with a greater profit margin, the movement of Chinese agricultural production from grains to horticultural crops, requiring higher amounts of chemical inputs, is likely to worsen the increasing trend in consumption of fertilisers and pesticides. It is, therefore, urgent to take effective and immediate action for the development of a horticultural sector able to adapt to environmental change.

9.5.2 Capacity Building

All demonstration projects implemented within the Sino-Italian co-operation projects provided, together with technology transfer, a full package of training activities tailored to the specific needs of local farmers and technicians and also involving academic institutions and private companies in an attempt to establish cross-sectoral partnerships. Particular emphasis has been given to scientific collaboration with academic institutes and research centres, which lead, as a consequence, to the development of research programmes. At the micro level, there is a need for China's higher education institutions to learn how to identify, develop, and implement research and extension programmes well adjusted to the global and domestic scenario. At the macro level, there is a need for China's policymakers to consult experts in order to formulate appropriate sustainable development strategies and policy. In this perspective, strengthening capacity and efficiency of the role of universities towards government, industry and market operators is of strategic importance for the future sustainable development of Chinese agriculture and the promotion of innovative 'green' technologies.

Joint research programmes have been established between Italian and Chinese universities, as a consequence of the technology transfer projects, with the aim of making Chinese scientists and technicians more acquainted with some of the modern techniques used in the demonstration projects. It is worth underlining, as an example, that the academic partnership between Italian and Chinese academic institutes, consolidated over 8 years of activities showed high capacity to convey human and financial resources from private and public sectors into the development of students' mobility programmes. In 2005, thanks to the exploitation of different sources of co-funding, the number of students and researchers visiting Italian firms and research institutes for study and research activities on sustainable agriculture within MSc and PhD programmes rose from a few to over 50 people, and the time they remained rose from a few weeks to 3 years.

9.5.3 Involvement of the Private Sector

The involvement of the private sector has been particularly important. Each project has been implemented on a participatory basis, stimulating the creation of broad partnerships involving all relevant stakeholders, from governmental agencies to non-government organisations (NGOs), from academic institutions to private companies. The model followed is that of Type II Partnership, which emerged from the 2002 Johannesburg World Summit on Sustainable Development as a means to promote a full integration of public and private sectors at large. Nevertheless, it has been quite difficult to plan for the involvement of the private sector in long-term programmes and give the joint projects a market perspective, facilitating the introduction and commercialisation of environmentally friendly innovative technologies in China. Due to the fragile Chinese regulatory framework on intellectual property protection, the involvement of Italian private companies is often limited to stand-alone contributions within each single project (e.g. field visits, lectures during seminars and training, short-term internships and technology procurement). Even though there is the possibility of benefiting by collaborating on a governmental programme like the Sino-Italian Cooperation Program for Environmental Protection, and so mutually benefiting both sides, the scepticism shown by private companies about the restrictive Chinese regulatory system represents a great barrier against effective technology transfer. The first signs of potential co-operation, fully involving Chinese and Italian private companies in the development of innovative technologies for agriculture, have been shown by the project 'Sustainable plant protection in respect of the environment: modern techniques for the control of plant pest and diseases of horticultural crops in China'. The project, co-financed by the Italian Ministry of Production Activities and the Italian Trade Commission with the aim to stimulate partnership between SMEs and universities, has opened the possibility of developing co-patents of biocontrol products for pest and disease control. Similar opportunities emerged in Chongming Island for the creation of agricultural waste composting facilities for the production of biogas and organic fertilisers.

9.6 Conclusions

There is a growing trend for projects linking agriculture to environmental protection and sustainable development to be registered during the recent years. Agriculture is no longer addressed as separate sector. On the contrary its deep interconnections to the societal, economical and environmental aspects of sustainable development project objectives are now recognised. There has been a shift from agriculture per se to agro-environment, intended as a complex dimension where food production 'internalises' the principles of environmental protection and sustainable development.

Xu et al. (2006) zoned China into nine regions and 22 sub-regions, depicting Chinese agriculture in the year 2000. They showed that 16 among the 31 provincial units in Mainland China have reached a level of sustainable development. Most of the projects concerned areas needing support. As the projects were fully integrated within a broader sustainable development programme (Clini et al. 2008), they were able to involve all stakeholders in both project preparation and implementation. In particular, government institutions, academic institutions, public research centres and private companies have always been partnered with the aim to create a long lasting network of local and international researchers and experts supporting the development and the adaptation of sustainable farming systems as well as the design of a new regulatory framework supporting the adoption of innovative technologies.

Important in all co-operation projects have been education, training and information activities to enable the actual transfer into practice of targeted sustainable agricultural technologies and practices. District workshops and seminars have been organised in order to maintain involvement of partners in project activities to inform them of progress as well as to informing stakeholders on the scientific, technical and economic feasibility of upcoming new techniques and systems.

The experience gained through the implementation of technology transfer projects in rural areas of China shows the strategic role that sustainable crop protection plays towards the promotion of sustainable development (Gullino et al. 2008).

In this regard the demonstration activities, the integrations of stakeholders and the international co-operation played a key role in achieving good results and have relevance to many other national bi- and multi-lateral projects.

Acknowledgements This work has been made possible by the funding of the many Agencies listed in Table 9.1.

References

Brown LR (2005) Reversing China's harvest decline. In: Earth Policy Institute (ed) Outgrowing the Earth. W. W. Norton, New York, pp 133–155

Cao Z, Dong D, Wang X, Han L, Gullino ML (2008) The reaction of three tomato rootstocks to *Meloidogyne incognita* in China. J Plant Pathol 90:S2.340

Cao A, Guo M, Cao Z, Zheng C, Gullino ML, Camponogara A, Minuto A (2002a) Sustainable practices for soil disinfestation: a project between Italy and China. Proceedings 2nd International Conference on sustainable agriculture for food, energy and industry, Beijing, 2:1492–1500

Cao Z, Yu Y, Chen G, Minuto A, Camponogara A, Gullino ML (2002b) Ecological studies on nematodes in alternative technologies to the use of methyl bromide in soil fumigation. Proceedings 2nd International Conference on sustainable agriculture for food, energy and industry, Beijing, 1:85–92

Clini C, Musu I, Gullino ML (eds) (2008) Sustainable development and environmental management. Springer, Dordrecht, The Netherlands, 487 p

Davis AR, Perkins-Veazie P, Sakata Y, Lòpez-Galarza S, Maroto JV, Lee S, Huh Y, Sun Z, Miguel A, King SR, Cohen R, Lee J (2008) Cucurbit grafting. Crit Rev Plant Sci 27:50–74

Dong D, Cao Z, Wang X, Hu J, Gullino ML (2007) Effect of nematode resistant rootstock on growth characteristics and yields of tomato. Acta Horticulturae Sinica 34:1305–1308

Dong X, Song S, Zhang X (2006) China's agricultural development: challenges and prospects. Ashgate, Aldershot, 311 p

Gullino ML, Camponogara A, Clini C, Yi L, Guanghui X, Xiaoling Y (2002) Sustainable agriculture for environment protection: a Sino-Italian Cooperation Program. Proceedings 2nd International Conference on Sustainable agriculture for food, energy and industry, Beijing, I:948–952

Gullino ML, Camponogara A, Capodagli N (2008) Sustainable agriculture for environment protection: cooperation between China and Italy. In: Clini C, Musu I, Gullino ML (eds) Sustainable development and environmental management. Springer, Dordecht, The Netherlands, pp 431–449

Gullino ML, Camponogara A, Capodagli N, Xiaoling Y, Clini C (2006) Sustainable agriculture for environment protection: cooperation between China and Italy. J Food Agric Environ 4:84–92

Gullino ML, Camponogara A, Gasparrini G, Rizzo V, Clini C, Garibaldi A (2003) Replacing methyl bromide for soil disinfestation. The Italian experience and implication for other countries. Plant Dis 87:1012–1021

Invest in China (2007) Status of farm, forest, herd and fish. www.fdi.gov.cn (website under the MOFCOM)

Lohmar B, Gale F (2008) Who will China feed? AmberWaves June 2008. www.ers.usda.gov

MOA (Ministry of Agriculture of China) (2007) 2007 China Agricultural Development Report. China Agriculture Press, Beijing

MOA (Ministry of Agriculture of China) (2006) The Eleventh Five-Years Plan of Agriculture. Abstract in English, full text in Chinese. www.fdi.gov.cn

OECD (2005) OECD Review for agricultural policies – China. OECD Publishing, Paris
OECD-FAO (2008) Agricultural outlook 2008–2017. OECD Publishing, Paris
Rozelle S, Huang J (2006) China's rural economy and the path to a modern industrial state. In: Dong X, Song S, Zhang X (eds) China's agricultural development. Challenges and prospects. Ashgate, Aldershot, pp 43–77
Scott G (1999) Polymers and the environment. Royal Society of Chemistry, Letchworth, UK 132 pp
Wang X, Chen Y, Cao Z, Dong D, Gullino ML (2008) Three alternatives to methyl bromide against root-knot nematode in China. J Plant Pathol 90:S2.346
Wen J (2008) Report on the work of the government. Report delivered at the First Session of the 11th National People's Congress, 5 Mar 2008
Xu X, Hou L, Lin H, Liu W (2006) Zoning of sustainable agricultural development in China. Agric Syst 87:38–62

Index

MIX
Papier aus verantwortungsvollen Quellen
Paper from responsible sources
FSC® C105338

If you have any concerns about our products,
you can contact us on
ProductSafety@springernature.com

In case Publisher is established outside the EU,
the EU authorized representative is:
Springer Nature Customer Service Center GmbH
Europaplatz 3, 69115 Heidelberg, Germany

Printed by Libri Plureos GmbH
in Hamburg, Germany